Energy, Biology, Climate Change

Also by Emil Morhardt

Global Climate Change and Natural Resources 2014
Global Climate Change and Natural Resources 2013
Ecological Consequences of Climate Change 2012
Global Climate Change and Natural Resources 2011
Ecological Consequences of Global Change 2011
Climate Change and Natural Resources 2010
Ecological Consequences of Global Climate Change: Summaries of the 2009 Scientific Literature
Global Climate Change and Natural Resources: Summaries of the 2007–2008 Scientific Literature
Biology of Global Change
Global Climate Change: Summaries of the 2006–2007 Scientific Literature
Research in Natural Resources Management
Clean Green and Read All Over
Research in Ecosystem Services
California Desert Flowers
Cannon and Slinkard Fire Recovery Study: A Photographic Flora

Energy, Biology, Climate Change

J. Emil Morhardt, Editor

CloudRipper Press

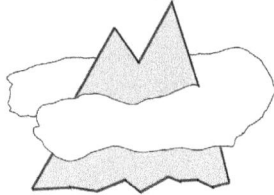

Cutting Edge Books

CloudRipper Press
Santa Barbara, California
www.CloudRipperPress.com

Morhardt, J. Emil
    Energy, Biology, Climate Change/ J.
    Emil Morhardt, Editor.

ISBN 978-0-9963536-0-1 (paper)

Table of Contents

Forward 9
J. Emil Morhardt

Section I—Energy 11

Energy Storage 13
J. Emil Morhardt

Wind Energy: Technology, Geography, Other Complications..... 27
Tim Storer

Solar Power: Impacts and Sustainability 45
Jincy Varughese

Hydraulic Fracturing in the United States 57
Alex Frumkin

Why Fracking Works—and Sometimes Doesn't........................... 71
J. Emil Morhardt

The Future of Tidal Energy... 87
Cassandra Burgess

Analysis of Nuclear Power as an Electricity Generation Option 103
Cameron Bernhardt

Section II—Biology and Ecology 115

Biological Responses to Climate Change 117
Anna Alquitela

Saving Endangered Species 129
Alexander Birk

Species Distribution Modelling and Climate Change................. 141
Kyle Jensen

Responses of Corals to Stress from Global Warming 157
Kimberly Coombs

Effects of Marine Conservation 177
Weronika Konwent

Effects of Climate Change on Agriculture 189
Adin Bonapart

Carbon Storage Capacity of Altered Forest Ecosystems 203
Stephen Johnson

Effects of Drought and Fire in Amazonian Rainforests 225
Maithili Joshi

Section III—Human Issues 239

The Social Cost of Carbon 241
Makari Krause

Climate Change and Urban Development 259
Dan McCabe

Anticipating Climate Change Impacts on Human Health 275
Amelia Hamiter

Climate Change and Human Health 295
Allison Hu

River Delta Systems and Climate Change 317
Rebecca Herrera

Natural Resources and Governance Policy: A Regional Survey 331
Lazaros M. K. Chalkias

Economic Impacts and Future Expectations of Global Warming
and Climate Change 345
Ali Siddiqui

About the Authors 357

Index 359

Forward

J. Emil Morhardt

The focus of this book is the interactions between energy, ecology, and climate change, as well as a few of the responses of humanity to these interactions. It is not a textbook, but a series of chapters discussing subtopics in which the authors were interested and wished to write about. The basic material is cutting-edge science; technical journal articles published within the last year, selected for their relevance and interest. Each author selected eight or so technical papers representing his or her view of the most interesting current research in the field, and wrote summaries of them in a journalistic style that is free of scientific jargon and understandable by lay readers. This is the sort of science writing that you might encounter in the New York Times, but concentrated in a way intended to give as broad an overview of the chapter topics as possible. None of this research will appear in textbooks for a few years, so there are not many ways that readers without access to a university library can get access to this information.

One place is scientific blogs on the Internet, and most of the material in this book will appear in the blogs ClimateVulture.com and EnergyVulture.com by mid-2016, but all of the material is available here.

This book is intended be browsed—choose a chapter topic you like and read the individual sections in any order; each is intended to be largely stand-alone. Reading all of them will give you considerable insight into what climate scientists concerned with energy, ecology, and human effects are up to, and the challenges they face in under-

J. Emil Morhardt

standing one of the most disruptive—if not very rapid—event in human history; anthropogenic climate change.

Section I—Energy

Energy Storage

J. Emil Morhardt

Wind and sunlight are intermittent, and thus is the energy derived from them. This is a problem for the electric grid; grid operators would like their energy to be "dispatchable" so they can dispatch it to meet demand. Wind and solar are inherently undispatchable—they come and go and ignore the grid operators needs. If only the energy they produce could be stored, even for a while, so that it was not necessary for the grid operators to maintain some type of "spinning reserve", typically a fossil-fuel-fired power plant burning fuel unproductively to keep its steam up and even its steam turbine spinning so that when called on, the turbine can turn the generator.

To some degree this problem is being ameliorated by the installation of the new generation of natural gas-fired turbines that can be up to speed in five minutes, but often the grid cannot wait five minutes to avoid a temporary brownout. Plus, the higher the penetration of intermittent renewables, the greater the problem becomes, so in order to gradually replace fossil fuel power plants with renewables, storage of energy produced by wind and solar will become mandatory.

Since the native output of photovoltaics is electricity, it would be ideal to store the energy in that form without having to convert it to some other form such as chemical potentials in batteries, or hydraulic head in pumped hydroelectric operations. Any time there is a conversion, the round trip inefficiency wastes electricity, and the loss can be quite significant.

Electricity can be stored as electric charges in capacitors, and increasingly "supercapacitors" are being experimented with for this purpose, being not only highly efficient, but charging and discharging rapidly, but they are expensive. If the conversion to some other type of energy is needed, flywheels are also potentially excellent, sharing the rapid charge and discharge and long lifetimes of supercapacitors, if not quite the round trip efficiency.

Lithium-ion batteries would, except for their expense, and slow rate of charge and discharge also be reasonably efficient. As discussed below, there is about to be an excellent source of cheap ones.

Wind turbines can as easily create compressed air as they can electricity, so one approach, discussed below, is to use the compressed air to drive turbines, thus the inefficiency is only one way rather than having to convert electricity to compressed air for storage then reconvert it to electricity, but that approach too is under development.

The native output of solar thermal power plants is heat, which can be stored temporarily in molten salt, delaying for a while the generation of steam to turn turbines. The entirely feasible delay of a few hours with very little loss of energy meets the needs of many grid operators who need only to store the energy from the heat of the day to the evening when the lights come on.

There are many other clever ideas in the works. More of them are discussed below.

Energy Storage Under Regulatory Uncertainty

A fundamental shortcoming of the US electricity grid is shortage of connected storage: the grid operators must instantaneously provide enough electricity to maintain an acceptable voltage and frequency by ramping generation up or down in real time, mostly using expensive CO_2-releasing electricity from natural gas peaking plants, and if there isn't enough demand to accommodate the electricity coming in from wind and solar, it just goes to waste. Amy L. Stein, writing in the Florida State University Law Review (2014), does a much more detailed job than I did in my brief introduction above of describing the existing energy storage facilities operating on the US grid, including hydroelectric pumped storage, compressed air energy storage, batteries, flywheels, and thermal energy, and the multiple grid services they provide. She then goes on to analyze the regulatory uncertainty that is partially responsible for the lack of grid storage, and makes an attempt at figuring out how to minimize it. This is a comprehensive document and worthy of reading by anyone interested in US energy storage initiatives.

Used Electric Vehicle Batteries for Home Energy Storage?

Why would you want to store electricity generated at home when you have a perfectly good connection to the grid and the power company buys back all the electricity your solar panels generate at the cost you pay for electricity? You wouldn't. But what if the power company started paying much less for the electricity you are shipping back than you could have purchased it from them—as they have been doing recently in Australia (Muenzel *et al.* 2014)—or decided to charge so much for transmission of electricity back to the grid that there was little point in selling it to them in the first place (as it appears they are contemplating doing in California from the radio spots I hear recently.) Now you might want a battery large enough to prevent any of the electricity you generate getting back to the grid, and ideally meeting all of your routine electricity demands. A cost-effective solution might be about to arrive just in time to offset these likely policy changes; used electric vehicle batteries.

New electric vehicle batteries store something on the order of 24 kWh—the Nisan Leaf, for example (Lacey *et al.* 2013). Around 2019 many of these will have degraded by 20% and need replacing. According to Lacey *et al.* , they ought to be perfectly usable for non-vehicular energy storage applications even when degraded to 50% of their original capacity, and according to Muenzel *et al.* , all you would need in a typical home is 6–7 kWh to obviate the need of using utility-provided electricity, so they would work fine in most cases with a payback period of 10 years, about the same as the solar panels feeding them. Thus, combining such a battery pack with existing or new residential photovoltaic installations is likely to make good sense about the time they become available.

By the way, doesn't it seem a little shocking that the amount of energy stored in a Nissan Leaf battery, which is good for something on the order of 40 miles, is enough electricity to run your house for three days? That makes it rather clear that improving gas mileage of cars ought to be far more effective at decreasing CO_2 releases than increasing household energy efficiency.

Used EV Batteries to Stabilize the Grid

If a single used electric vehicle (EV) battery, still functional but with it's storage capacity degraded by 20%, could be used to store energy from photovoltaic panels at home (see Dec 30 post), a bank of them could be used to stabilize the whole grid, according to Gillian Lacey and colleagues (2013) at Northumbria University in the UK. Their particular emphasis is "peak shaving"—supplying enough electricity at times of peak demand that additional fossil-fuel generation can remained turned off. This would also help regulate the line voltage and allow some "upgrade deferral"—putting off investing in needed new or more efficient sources of energy. A particular value of this type of storage is that it would be at the low voltage end of the distribution system, closer to the end user, thus decreasing the potential line losses that would occur if peak power were supplied by regular generation systems, and increasing the life of transformers which are particularly stressed under peak loads. It might also mean that this type of storage would be "distributed"—maybe one battery per residential transformer. Since most large-scale current photovoltaic generation stations generate at low voltage, used EV batteries might be useful there as well.

In grid applications, the authors have assumed only a single charge-discharge cycle per day (matching the daily peak—probably exactly what would occur at home PV installations as well), which would be effective in prolonging the life of the batteries.

Beefing up a Wind Turbine with Compressed Air

If your wind turbine isn't going fast enough to meet the demands of the grid, blow on it a little harder: that's the general idea suggested by Sun *et al.* (2014). The concept is a little like a hybrid electric vehicle; if the internal combustion engine isn't going fast enough, give it a little boost from the electric motor connected to it. Except in this case, it's that if the wind turbine isn't going fast enough, goose it with a little compressed air. You might be envisioning a compressed air nozzle pointed at the turbine blades, but there's a better way: use a motor driven by compressed air to speed up the turbine. One novel aspect to this study is that the device envisioned as a compressed-air motor is something called a scroll expander, or scroll-type air motor, a new type of pneumatic drive, but that doesn't seem to be central to the idea—any suitable air-driven motor should work. The main point is to have it integrated with the wind turbine so that when needed, it can help out in the short term.

To function, it needs a source of compressed air, but where that might come from isn't dealt with in the paper. It could be generated by the same wind turbine using excess electricity to run an air compressor, but maybe there are other sources as well, and I think the authors envision a centralized source of it—rather than one compressor per wind turbine, one compressor per field of turbines, all connected with air hoses, and maybe that one a wind turbine whose native output is compressed air rather than electricity. The authors created a test rig in the lab that simulated all aspect of the system, and conclude that 55% of the energy used to compress the air could be expected to be returned from the boosted wind turbine, which is fairly good for pneumatic motors, and the wind turbine itself was much better at providing steady reliable power with the compressed air device than without.

Compressed Air Hybrid Vehicles?

The usual candidate power supplies for the non-fossil-fuel part of hybrid vehicles are chemical batteries, supercapacitors, and flywheels, all powered up using electricity, and generating electricity when their power can usefully replace or supplant the main power source, the internal combustion engine. But these types of electrical storage and the motor/generators they utilize are complex, sophisticated, and ex-

17

pensive, and have barely appeared at all in the developing parts of the world where fossil fuel use is growing fastest. Maybe there is a simpler, cheaper option. One possibility is compressed air energy storage. All you need is a tank (cheap), a reversible compressor (fairly cheap), and a way to link it to the engine. That last part is tricky because the general run of such systems work optimally at a specific pressure, but their performance falls off dramatically as pressures in the tank exceed or fall below optimum as would be expected as the tank is being pressurized or depressurized. The simple solution, according to Brown *et al.* (2014), is to use inexpensive check valves on the tank to prevent over-compression and over-expansion, and an infinitely variable transmission between the compressor and the engine that can operate efficiently at a range of tank pressures. The transmission adjusts by changing the number of thermodynamic cycles of the compressor executes per driveshaft rotation.

Nobody is doing this yet in road vehicles, the authors believe, because the standard approach uses the engine as a compressor and requires the addition of variable valve timing with expensive actuators, rather than the relatively inexpensive addition of a dedicated external compressor and transmission. Although the efficiency of the proof of concept system they cobbled together is not very high—only about 10% of the energy converted to compressed air comes back to drive the engine—the authors figure that if the exhaust heat of the engine were used to keep the air tank hot (which would cost nothing in terms of energy) a round trip efficiency of 47% might be realized at a cost lower than that of battery hybrids, and in a package that would last a long time and could be repaired using mechanical capabilities common in the developing world.

Trackside Flywheel Energy Storage in Light Rail Systems

Light rail systems, like hybrid electric vehicles, use their electric engines to generate electricity when they are slowing down, a process called regenerative braking. In hybrid electric vehicles, the energy usually gets stored in lithium-ion batteries, which work well because they are comparatively light-weight and not overly bulky. If neither of these were constraints, then flywheels or supercapacitors would be a better choice because they can deliver power faster and they take much longer to wear out. Of the two, flywheels are lighter, less bulky, cost less, and have longer lives according to a study by the UK Rail and Safety Standards board (Kadim 2009). If they are installed alongside the tracks rather than on the trains, weight and bulk are not very important but cost and lifetimes still favor flywheels.

Gee and Dunn (2014) set out to estimate how much energy could be saved in an electric light rail system by using flywheel energy storage alongside the tracks; the result is 21.6% assuming the current mix of regenerative and friction braking, with even more savings if friction braking were reduced further. Additional benefits would include a 30% reduction in substation peak power, thus a smaller investment in substations would be required on new installations. The authors didn't do an economic analysis, but cite another paper that suggests a payback period for installing such a system of only five years. What are they waiting for?

Off the Grid, Batteries Not Included

If there's a need for electricity, but there aren't any power lines nearby, the approach of choice today, in sunny climes, is photovoltaic (PV) panels connected to batteries. But the total amount of electricity that can be stored is then dependent on the number batteries, and if a relatively large amount of storage is needed, this could be prohibitively expensive, heavy, and not very portable. With the advent of hydrogen fuel cells small enough to fit into an automobile and operational at low temperatures, perhaps a fully self-contained electrical generation system could be based on PV panels electrolyzing water to make hydrogen gas, which could then be stored in low-pressure tanks in amounts as large as needed. That's what Cabezas *et al.* (2014) decided to experiment with.

The impetus was that in their country, Argentina, there is much need for off-the-grid electricity in isolated communities, military outposts, and mountain huts. The authors put together the system using commercially available components and their own electrolyzer (in which the electricity from the PV panels splits water into hydrogen and oxygen gas) and fuel cell stack (which recombines the hydrogen and oxygen to produce electricity). The gases were stored at a pressure only slightly higher than atmospheric pressure, ruling out the need for expensive compressors or high-pressure tanks.

The whole system was light-weight, easily transportable, and inexpensive, and seems to be a good candidate for development into a unified module that could be purchased off-the-shelf for off-the-grid needs.

The authors suggest that their system has the advantage of not incurring the energy losses involved in charging and discharging batteries but they did not do an analysis that directly compared the cost or efficiency of their system to a reasonable battery storage alternative.

Flywheel Versus Supercapacitor for Running a Small Electric Ferry

When we think of all-electric cars, we think lithium-ion batteries because they are lightweight and have a high power density. For ships, light-weight doesn't matter so much, and it turns out there are types of shipping routes that don't need very much energy storage: think ferries, specifically the plug-in ferry Ar Vag Tredan (the "electric boat" in Breton), a zero-emission passenger ferry crossing the Lorient roadstead 56 times a day. When parked between trips it can recharge its supercapacitor more-or-less instantly (that's a main feature of supercapacitors—that and their ability to discharge their power equally quickly to meet any need for power the ship may have.) How would a flywheel energy storage system work compared to the existing supercapacitor? That's the question asked in a new paper by Olivier *et al.* (2014).

The answer is, pretty well. This is a full-on engineering paper stuffed to the gills with calculations, looking at the problem every which way (that's what engineers do), but they are still looking at a basic system: optimizing the flywheel, the electrical machine (they mean motor I bet) and the power converter which converts electricity into flywheel spinning and vice versa. They assume the flywheel will last 20 years, the ship crossing 35 times a day (less than it actually does it looks like), spending ten minutes motoring across—expending 16 kWh—and five minutes at each end loading, unloading, and re-charging.

Wearable Supercapacitors: Making Devices More Flexible

Maybe someday you will be able to recharge your gadgets by plugging them into your jacket, which you charged up in a few seconds from a convenient wall plug or maybe even from wi-fi. Yu *et al.* (2014), at the School of Chemistry and Chemical Engineering, Nanjing University, in China, have fabricated experimental sheets of flexible layered conductive and non-conductive materials that they envision as eventually wearable. We all get tired of waiting around for batteries to charge, but supercapacitors charge almost instantly. They don't usually have much energy storage capacity though—you don't get as much energy storage per unit weight or volume as you presently can from batteries—but if they are built into something that you need to carry around with you anyway, that might not be so important. A good example is the plug-in electric boat. Boats don't care much about how large or heavy something is, but they need to be fueled rapidly. So if you could store all the energy you need quickly in your jacket, your battery-powered devices could recharge in your pocket, wherever you are.

Yu *et al.* 's flexible supercapacitors, however, have very good energy storage capacity (50–60 Watt hours per kilogram), as good as the lead-acid batteries in your car, but unlike your car battery, they are quite thin and flexible.

Hybrid Energy Storage for CubeSats

CubeSats are cool. No, actually very cold, since they're out in space. But they are reproducing like rabbits. There are well over 200 of these little 10 cm x 10 cm x 10 cm cube satellites have been launched into orbit by tucking them into the nooks and crannies in the launch vehicles around much larger satellites. (Some are multiples of cubes, 10 cm x 20 cm, or 30 cm.) They need energy. Until now they have been powered in the main by lithium ion batteries like those in your computer, and charged by the photovoltaic panels that make up a CubeSat's skin. The thing is that these batteries don't work very well when they are cold; the speed of electrochemical reactions, just like those of every other chemical reaction, are modulated by temperature—the colder the slower. The current Li-ion batteries don't work at all below –10°C, yet CubeSats headed for deep space are expected to encounter temperatures of –40°C some of the time. So if you have a CubeSat process that needs power at low temperatures or a short-term burst of power faster than the batteries can provide, you need help.

Supercapacitors, even though they don't hold as much charge per volume as a lithium ion battery, don't much care what the temperature is, plus they charge and discharge much faster than batteries. It makes sense then that some enterprising engineers would combine the two systems to get the benefits of both; the engineers in question are at the CalTech Jet Propulsion Laboratory and at Cal State University, Northridge, nearby. They have just completed a study demonstrating that this hybrid system will solve the problem. The general idea is similar to combining supercapacitors with battery banks to help windpower installations integrate better with the grid but in this case the system is *much* smaller and in addition to delivering power rapidly without damaging the batteries, it operates at temperatures lower than any wind farm is likely encounter.

Supercapacitors save Windpower Batteries

Windpower, because it is intermittent, works best on the electrical grid if it has some energy-storage facility connected to it. Batteries are the simplest approach, and the low cost of lead-acid batteries makes them good candidates, but they resent being randomly charged and discharged (especially deeply discharged) at the will of the wind, and die prematurely. Enter the supercapacitor; it can be charged by wind turbines much faster than a battery, can deliver its stored energy to the grid much faster as well, and doesn't resent it at all, even being deeply discharged at every cycle. Engineers at Kocaeli University connected a supercapacitor in parallel with a battery so that it would buffer transient current surges, saving the battery to do what it does best. The system worked just like one might expect. There are some graphs in the paper showing just how the current flowed, and I think it is a nice example of an experimental setup to look into these types of hybrid energy storage systems.

Storing Wind Energy on Islands is Risky, Economically

Electrical storage is needed to meet demand when solar and wind power become intermittent, particularly in off-grid systems like isolated oceanic islands. Both fast-response and long-term storage is needed; fast response (seconds to minutes) systems to prevent grid overload during short periods of high load (everybody on the island turns on the air conditioners at once), or of clouds passing over sun, or the wind stopping for a while. Longer-term storage (hours to days) can shift the time the power from renewables is produced to when it is really needed, or is at its highest value (more valuable than, say, the diesel generator sets that the island was previously dependent on. But at what cost? Moazeni *et al.* (2014) modeled the expected revenues to be gained in an isolated island environment of intermittent wind, shifting electrical loads, and a pumped storage hydroelectric system station. It turns out that optimizing the amount of storage and using it in the most economically efficient manner is no trivial matter, with the economic returns deviating as much as 29% from what might have been expected by the investors, depending on how the system is managed. What such systems operators apparently clearly need is electrical engineering modelers like these authors to figure it out for them.

Energy Stored in the Wire?

How about storing the excess energy not chemically in batteries, not as potential energy in pumped hydroelectric reservoirs, not as compressed air, or heat, but just as excess charges? That's what capacitors do, and supercapacitors are widely used for storing not-too-much energy for immediate release when needed. What if the electrical cables themselves could be turned into supercapacitors. Yu and Thomas (2014) figured out how to do just that by using nanowires spaced around the central charge carrying cable to accumulate the excess charge. If wires like this become economic, they could be used to connect photovoltaic panels and wind turbines to the grid, automatically leveling out the delivery of power when the source becomes intermittent.

Fast Discharge Batteries: Electric Eels

Finally, though this has more to do with generating electricity and using it wisely than storing it *per se*, one of the oldest electric batteries is the 600 Volt biological battery of the electric eel *Electrophorus electricus*. These animals search about for hidden prey by emitting a couple of high voltage pulses from time to time to see if anything jumps. If it does, they sidle on over and cut loose a volley of high frequency (~400 Hz) pulses that the muscles of the prey apparently interpret as coming from their own nervous system. The result is many muscles contracting simultaneously, paralyzing the prey into a state of whole-body muscle contraction known as tetanus (similar to the eponymous disease) and the eel sucks them in. This all happens pretty fast, on the order of milliseconds. If the eel fails to suck them in they often just swim away.

In a series of elegant experiments, Kenneth Catania (2014) at Vanderbilt University explored this in some detail by isolating immobilized fish behind an electrically-transparent barrier—keeping the eel at bay—while he measured the eel's volleys and the prey's responses. Given that everyone knows electric eels exist, and that electrophysiologists can't help being interested in them, it is somewhat surprising that all this wasn't known before. Fisheries biologists know all about the effect of shocking fish because they use electrofishing gear to temporarily stun fish so they can be captured and counted, and I had always assumed it was doing something similar to a hunting electric eel. Dr. Catania and the editors of science interpret this phenomenon in a more nuanced way, characterizing the eels' behavior as "…remotely control[ling the] muscles of their prey." Somehow, viewed in that light, the process seems far more nefarious than simply stunning the prey.

Conclusions

Cheap, efficient storage of the energy produced from renewables is increasingly needed to stabilize the electric grid. The only such storage even moderately widely deployed at the utility scale at the moment is pumped hydroelectric storage. This works well, but is highly inefficient, wasting much of the energy that could have been stored, and in its present configurations requires large open reservoirs uphill and downhill from the power plant, so side from needing the right topography, there are many environmental issues.

Instead, inventors are working on probably hundreds of schemes that would be less expensive, more efficient, and less problematic to install. If a really good one is discovered, there is considerable money to be made.

References Cited

Brown, T., Atluri, V., Schmiedeler, J., 2014. A low-cost hybrid drivetrain concept based on compressed air energy storage. Applied Energy 134, 477-489.

Cabezas MD, *et al.* , Hydrogen energy vector: Demonstration pilot plant with minimal peripheral equipment, International Journal of Hydrogen Energy (2014),

Catania, K., 2014. The shocking predatory strike of the electric eel. Science 346, 1231-1234.

Chin, K., Smart, M., Brandon, E., Bolotin, G., Palmer, N., Katz, S., Flynn, J., 2014. Li-ion battery and super-capacitor Hybrid energy system for low temperature SmallSat applications. SSC-14-VII-9, 28th Annual AIAA/USU Conference on Small Satellites.

Erhn, K., Aktas, A., Ozdemir, E., 2014. Analysis of a Hybrid Energy Storage System Composed from Battery and Ultra-capacitor, 7th International Ege Energy Symposium & Exhibition, June 18-20, 2014, Usak, Turkey

Gee, A., Dunn, R., 2014. Analysis of Trackside Flywheel Energy Storage in Light Rail Systems. IEEE Transactions on Vehicular Technology DOI 10.1109/TVT.2014.2361865

Kadhim, R. 2009. "Energy storage systems for railway applications. Phase 1 Report.," Rail and Safety Standards Board Report, T779, Interfleet Technology Ltd., Derby, UK, Sept. 2009.

Lacey, G., Putrus, G., Salim, A., 2013. The use of second life electric vehicle batteries for grid support, EUROCON, 2013 IEEE. IEEE, pp. 1255-1261.

Moazeni, S., Powell, W.B., Hajimiragha, A.H., 2014 (in press). Mean-Conditional Value-at-Risk Optimal Energy Storage Operation in the Presence of Transaction Costs. IEEE TRANSACTIONS ON POWER SYSTEMS

Muenzel, V., de Hoog, J., Mareels, I., Vishwanath, A., Kalyanaraman, S., Gort, A., PV Generation and Demand Mismatch: Evaluating the Potential of Residential Storage. http://www.juliandehoog.com/publications/2015_ISGT_PotentialStorage.pdf

Olivier, J.-C., Bernard, N., Trieste, S., Mendoza, L., Bourguet, S., 2014. Techno-economic Optimization of Flywheel Storage System in transportation, Symposium de Génie Électrique 2014.

Stein, A., 2014 Reconsidering Regulatory Uncertainty: Making a case for energy storage. Florida State University Law Review 41, 697.

Sun, H., Luo, X., Wang, J., 2014. Feasibility study of a hybrid wind turbine system–Integration with compressed air energy storage. Applied Energy DOI: 10.1016/j.apenergy.2014.06.083.

Yu, C., Ma, P., Zhou, X., Wang, A., Qian, T., Wu, S., Chen, Q., 2014. All-solid-state flexible supercapacitors based on highly dispersed polypyrrole nanowire and reduced graphene oxide composites. ACS Applied Materials & Interfaces.

Yury Gogotsi writes about it in the 29 May 2014 issue of Nature.

Wind Energy: Technology, Geography, Other Complications

Tim Storer

As the threat of climate change continues full bore, governments, private businesses and citizens all seek alternate forms of energy. Ideally, these sources will be both renewable, and have comparatively lower emissions than traditional fossil fuel sources, which contribute roughly 69% percent of total anthropogenic carbon emissions per year (IEA, 2014). Wind energy offers an excellent example of such a power source (Mostafaeipour 2010). It is renewable, and as is illustrated in the final portion of this chapter, relatively emissions-free (Nugent and Sovacool 2014). Largely due to annual decreases in prices, wind energy production continues to rise globally (Sahin 2004). In 2013, the most recent year available, 318,596 MW of wind power was operational globally (GWEC 2014).

While wind energy is renewable and reliable in the long term, it is extremely volatile in the near term. Highly variable power supply makes wind power plants unfavorable in an electrical grid compared to dispatchable energy sources, such as coal or natural gas. In order to better address this issue, several different approaches have been taken, such as partnering wind energy with reliable sources, including energy storage systems (ESS) and increasing the accuracy of wind speed predictions. Each of these options is discussed in order in the first three sections of this chapter.

Several factors must be considered when planning a wind energy project. Many potential locations are dismissed for inadequate

27

average wind speeds, and many more are dismissed due to land-use conflicts. For example, the installation of terrestrial turbines often involves clearing land, destroying local species habitat, and creating an aesthetic nuisance. These operations must also be subject to economic analysis indicating under what circumstances they could provide power at a competitive price. To illustrate the complex issues associated with determining proper sites for harvesting wind energy, this chapter includes recent studies on wind potential in Norway, Thailand, and Iran.

Wind power operations are not without their share of environmental hazards. Numerous studies have documented the adverse effects of wind turbines on flying animals, but most of this research has focused on birds (Telleria, 2009). This chapter includes a recent study on adverse turbine effects on migratory bat species. In addition to harms against local environments, wind turbines also contribute a share of GHG emissions to global climate change. Wind greenhouse gas emissions come mostly from construction and raw material extraction, and the total emissions are tiny in comparison to gas or oil. That said, all these adverse effects should be considered when planning a wind energy project, especially since the environmental sensibility is often touted as a premier reason for wind energy development.

Overall, this chapter seeks to do two things: first, to highlight the recent and exciting literature on wind energy in the last year. Second, it hopes to provide an overview of the state of wind energy in general, and give a sense of the challenges of implementing it.

A Convenient Partnership Between Carbon Capture and Wind Energy

Carbon Capture Storage (CCS) technologies help to reduce emissions from fossil fuel energy operations, such as coal fired power plants. While these technologies have the benefit of reducing greenhouse gas emissions and making the operations more climate friendly, they are costly for extraction companies. Wind power has the benefit of low emissions, but is dependent on weather and fails to provide a stable energy supply. This paper identifies a way to reduce the cost of CCS, which involves partnering with wind powered energy. Bandyopadhyay and Patiño-Escheverri (2014) find that this partnership can make CCS vastly cheaper for the producers and the partnership would also create additional incentives for developing renewable energy sources in the form of wind power. Through the partnership, power providers will have the flexibility to direct power to multiple uses depending on price fluctuations, thus minimizing profit loss from incorporating CCS.

The paper examines a form of carbon storage that uses " CO_2-rich" amine solution. One downside of this form of CCS is that it requires energy to break down the amine solution, and this can decrease net power output from the coal plants by 20-40%. Needing to divert so much power to CCS is one of the major drawbacks preventing power producers from including CCS in their operations. This paper explains that one major way to reduce the cost of diverting power towards breaking down the captured carbon products is to store the power and not dispose of it immediately. By storing the CO_2-laden amine solution, the companies can choose to break down the carbon products during times of low electricity prices, and thus time their CCS breakdown with times of low profit. Overall, this helps minimize cost by allowing them to maximize their profits and only direct energy towards CCS when profits would be low.

This effect is further helped by the inclusion of wind energy systems. Wind power from excessively windy days –that might otherwise overwhelm their transmission lines– can instead be channeled to breaking down of the captured carbon compounds. This essentially allows the stored carbon compounds to serve as storage for the wind energy, which usually suffers the disadvantage of not being easily storable.

Overall this partnership offers a clever way to mitigate the inconsistency of wind power and the costs of CCS on coal fired power plants. By comparing CO_2 capture (CoC) cost estimates and Levelized Cost of Electricity (LCOE) in the Chicago area, Bandyopadhyay and Patiño-Escheverri were able to compare the relative costs of different energy configurations involving coal plants and even the option of replacing them with gas-fired ones. They found that hybrid plants had significant decreases in CoC and in LCOE as compared to continuously operating coal plants, thus making CCS more attractive than previously thought and offering potential for lower energy prices. Under the price fluctuations observed in the Chicago Hub, hybrid plants were favorable in about 70% of the wind energy sites they considered.

Additionally, these findings suggest hybrid power sources as a further incentive for developing wind technology, and may be a better motivator than power subsidies for wind farms. Lastly, they find the hybrid systems to be more cost effective in the face of CCS than new natural gas plants. The exact cost reductions vary according to the assumptions, but all benefits are likely to be understated given that the study didn't account for cost reductions from avoiding new transmission line construction in a case where wind and coal plants were developed separately. The study also assumed that hybrid operations would adjust on an hourly basis, when they could probably operate much faster. This would allow for even more cost-effective power generation.

Currently, wind-coal hybrid partnerships are not in place, nor are any CCS technologies. Reducing costs to CCS are a worthwhile step towards making them economically (and politically) viable, and ultimately, operational.

Of the many Energy Storage Systems, Integrated Hydrogen-Oxygen Storage Stands Out

Wind power comes with the disadvantage of intermittent gaps in energy production and instances of excess supply. This variability puts strain on the electric grid and is the primary barrier to large-scale wind power integration. In order to combat this issue, various forms of energy storage have been considered to bridge the gap between supply and demand of wind power. Gao *et al.* 2014 conduct a brief literature review on all existing energy storage systems (ESS) for wind power. Each method comes with drawbacks associated with scale, cost, or safety, but hydrogen-oxygen storage was seen here as the best future option. By improving storage technologies, wind energy will become more viable in the market and help to reduce the share of energy coming from fossil fuels that contribute to climate change. In addition to the literature review, this study examined a possible hydrogen-oxygen ESS in Jiangsu Province, China and saw that such an operation could be profitable in the current market.

While there are some operational forms of ESS, there is a variety of issues preventing ESS –and subsequently, wind power– from becoming widespread energy sources. For example, battery power is too costly and difficult to build at a large scale, systems that involve pumping water upward for energy storage have geographical limitations, and magnetic energy storage has low storage time. In the case of hydrogen generation from electrolysis, the costs are simply too high to be competitive in the energy market with capital costs of 1000-2500$/kW (when they need to be near 400 $/kW).

Hydrogen-oxygen combined storage consists of electrolyzers that break water down into hydrogen and oxygen. The hydrogen and oxygen are combusted to form super-heated steam that powers turbines. The system is closed, and uses water as a recycled fuel. Gao and colleagues examined three variants of hydrogen-oxygen ESS: simple integrated ESS, integrated ESS with a feed water heater, and an integrated ESS with both a feed water heater and a steam reheater. In simple terms, these systems each contain an additional measure to capture heat from the steam turbines and use that heat elsewhere in the process, thus improving efficiency. All of these integrated systems contain a complex web of mechanisms that can be adjusted alongside price fluctuations in the power market to minimize costs. The former two had roughly equivalent efficiencies of 49%, but the latter system had efficiency of up to 54.6%, thus demonstrating the benefits of feed water heaters and steam reheaters.

While the 54.6% efficiency of the fully integrated system is marginally below that of some other ESS technologies, hydrogen-oxygen systems come with certain advantages. They can be implemented on a large scale, are fully eco-friendly, not limited by geographical and material restraints, and can be adjusted rapidly based on demand changes. The system was analyzed under two extreme scenarios: an "intermittent operation mode" simulating an extremely variable wind supply, and "continuous operation mode" simulating a perfectly steady supply. Because of how effectively the system dealt with times of low wind, it was actually more profitable under the intermittent scenario with annual income of $13 million per year. Real wind conditions lie somewhere between these extremes, and efficiencies of approximately 50% and prices of 0.03–0.05$/kWh were estimated.

Accurate Estimation of Wind Energy Storage Needs Requires an Autocorrelated Approach

Variation in wind energy output puts strains on the electric grid and often discredits otherwise viable wind operations from being competitive in the electricity market. This is especially true in smaller islands (with smaller electric grids) and in areas with legal restrictions on the percentage of energy coming from uncertain sources. Energy storage systems (ESS) are often implemented to mitigate power variation, and calibrating the size of the storage system involves comparing the installation and operation costs with the degree of expected wind variation. Previous models for anticipating wind variation have statistical flaws that result in undervaluing the importance of ESS. Haessig *et al.* (2015) examine the inclusion of an "autocorrelated" approach to modeling wind predictions that is better suited for determining the proper ESS size for a given wind operation. They conclude that their approach is an improvement over previous models and can accurately recreate existing data.

Haessig *et al.* critique existing wind variation models in two ways: some models fail to account for the correlation between individual variations, and others require too many variables and are therefore difficult to use in practice where there are limited data. The goal was to design a statistical approach that could be readily used and would also provide more accurate representation of real wind volatility. Both issues were addressed, starting with improvements in the ability to account for the correlated variations. In simpler terms, Haessig and colleagues wanted to account for the fact that hourly predicted wind speeds throughout a given day should not be independently calculated because they are often interrelated. Therefore, the study examines a model that incorporates the speeds in a given hour into the prediction of the following one, with the weight of a given hour declining exponentially over time in predicting the following hours. This was done with a Monte Carlo simulation that incorporated the interconnectivity of individual variations in wind speed.

It is openly admitted that this model does not account for all types of wind variability, and is designed for one use only: predicting the appropriate size of an ESS at a given wind power location. In that respect, the model works very well. It was able to accurately reproduce actual wind variation data with a correlation coefficient of 0.8–0.9% depending on which data were used. Improvements in these estimates will allow future wind operations to be more carefully calibrated, have more ideal ESS size, and therefore be better suited for inclusion in the electric grid. This study offers an example of a statistical improvement that ought to improve the standing of wind energy in a competitive energy market, thus lowering the amount of fossil fuel emissions from energy production. This is especially true considering that larger ESS will lower the frequency with which wind operations fail to provide any energy, and electricity must be taken from "spinning reserves" —electrical sources that can be quickly amped up to account for gaps in energy supply. Spinning reserves are usually fossil fuels, so this is yet another way that overall emissions could be lowered from improvements in ESS design.

Norway's Path to Zero Emissions: Large Scale Hydrogen Production from Off-Grid Renewable Sources

Norway currently generates over 95% of its power from hydroelectric dams, making it one of the most climate friendly energy systems on the planet. In efforts to bring Norway carbon neutral by 2050, the government aims to eliminate emissions from the transportation sector. Konrad Meier of the Stuttgart University of Applied Sciences examines the possibility of using a hypothetical 100 megawatt offshore wind farm to generate hydrogen fuel via electrolysis. Because water hydrolysis uses only electricity and water, it offers an emissions-free means to generate hydrogen as long as the electricity is generated from a renewable source, such as wind power. This could achieve Norwegian political goals of carbon neutrality by providing the hydrogen necessary to transform their transportation sector. Unlike other proposed wind-to-hydrogen technologies, Meier examines an off-grid operation, rather than producing hydrogen at the fuel refill site. The analysis was conducted under three scenarios, and the hydrogen from this proposed operation is profitable in the energy market under only the "best case" scenario.

This is a clever use of wind power for several reasons. First, if this operation were integrated into the power grid, wind variability would become an issue. Keeping the production off-grid avoids costs of transmission infrastructure and variable supply. Second, the variability also makes exporting excess power to the E.U. infeasible. The remaining 5% of domestic power is more likely to come from untapped hydroelectric resources, so wind has no use in Norway either. Using wind power for hydrogen synthesis circumvents these issues that have previously prevented wind production from being a viable energy source in Norway.

To examine costs, Meier used a location proximal to an operational German wind farm, Alpha-Ventus, and incorporated its data. He uses 2010 data as his "worst case" scenario, which is a very conservative baseline considering how much lower the power output had been in previous years and how the proposed system would not be subject to transmission losses. The "best case" scenario was calculated simply by the predicted estimates. Given the likely increases in electrolysis efficiency in upcoming years, this scenario also yields conservative estimates of overall output costs. Meier discusses four types of electrolysis, but focuses on proton exchange membrane electrolysis cell (PEMEC) and solid oxide electrolysis cell (SOEC) that require water as the only input material. Unfortunately, research on the efficiency of PEMEC and SOEC is unable to offer precise estimates, and herein lies a major source of ambiguity in the study.

Because there is currently no market for hydrogen transportation fuel, this study is limited by the assumption of a future in which infrastructure has been implemented to support a hydrogen market. Unfortunately, given how variable the results are (dependent on optimistic/pessimistic assumptions), it is unclear whether such an investment is worthwhile at all. However, in as much as the best and worst case scenarios were estimated in a very conservative way, it is possible that such an operation could be an economic way to transform the transportation sector. Further research is needed on the efficiency of the PEMEC and SOEC processes to indicate whether the proposed wind farm is an economically viable solution to attaining a carbon neutral transportation sector.

Thailand Has the Necessary Wind Conditions to Reach Renewable Energy Goals

Many developing economies are undergoing an energy transformation, and in the face of global warming, there has been a push towards investment in renewable sources, such as wind power. Chingulpitak and Wongwises (2014) review the current status of wind energy development in Thailand. The Thai government has stated goals of increasing its use of renewable fuels to 25% by 2021, and wind energy is a large component of this transformation. In 2012, only 111.7 MW of wind power was generated, but the Thai government aims to increase production to 1800 MW in this timeframe. In addition to its own worth, Thailand's energy transformation can provide insight into the challenges of other developing nations around the globe.

Like many developing countries, Thailand's total energy consumption is rising fast, with a 13% increase from 2007–2012. It currently is the 25th biggest electricity user globally, but is the largest consumer in Southeast Asia. This time period also coincided with minor shifts away from coal and oil towards renewable energy and natural gas. While these shifts are in the desired direction, they are not on pace for the 2021 goals.

Wind power operations have been steadily introduced to Thailand since the mid 1980s, and Thailand currently hosts two large-scale wind operations and many smaller ones. As part of the Thai policy to promote renewable energy, these companies receive energy subsidies of 0.08–0.11 $/kWh over ten year periods. In addition to current operations, many potential sites exist.

The coastal area of the Thai Gulf has the highest potential for wind energy. Potential sites must have wind speeds of more than 6.4–7.0 m/s at 50 meters of elevation. Because smaller wind turbines of less than 30 meters can offer the benefit of producing off-grid energy, they can be a successful provider to remote rural areas. Unfortunately, only 9% of rural populations live nearby potential small wind power sites in Thailand, lower than that of Laos (13%) and Vietnam (30%), thus limiting the potential benefits to rural communities. However, Thailand does have additional potential for larger scale wind operations, with multiple turbines over 65 meters tall. Multiple suitable areas, such as Killom, Monlan and Maehae, are ready for large-scale wind operations capable of producing 18.7–44.9 GWh/yr.

There are three primary obstacles to development: conflict with other land uses, production costs due to lacking infrastructure, and conservation concerns from development. Thailand has the physical capabilities for its desired wind production, but these policy obstacles must be tackled with thoughtful negotiation.

Potential for Additional Wind Energy in Major Iranian Cities, Tabriz and Arbadil.

Threats from climate change and unreliable fossil fuels prices have spurred initiatives to develop renewable energy in many parts of the world. In the case of Iran, wind constitutes the biggest share of potential renewable energy growth. Previous research indicates that Iran has 18,000 MW worth of potential economically competitive wind energy as compared to the 109 MW currently in use. In efforts to determine the potential for wind energy development in two

Iranian cities, Tabriz and Ardabil, Fazelpour *et al.* (2015) analyze existing wind speed data, turbine efficiencies, and local Iranian electricity prices. It is found that even a wind turbine of 25 kW, a small scale for industrial turbines, could provide cost-competitive electricity in both locations for most months of the year. This research contributes to a growing body of literature highlighting potential wind sites, and an overall trend in an increased share of global energy coming from renewable sources.

A Weibull distribution method was selected to analyze wind speeds due to its simplicity and previous usage in the literature. Data from wind speeds at a height of 10 meters above ground were collected at three hour intervals on all days between 2005–2010. This distribution model emphasizes the variability of wind by showing the frequency of various speeds in a continuous bell graph. Special attention was paid to the variables k and c, which represent how "peaked" the distribution is, and the maximum windiness, respectively. Related to c is the variable P/A, or wind power density. This describes the total power available from a turbine of a given height in that location.

In Tabriz, the most populous Iranian city, results show promising wind potential. The highest average wind speeds of 5.71 m/s occurred in the month of July, with the lowest speeds of 2.14 m/s in September. The largest deviations in wind speed were observed in August (2.87 m/s), and the smallest in December (1.48 m/s). The highest monthly power densities were in summer months, with values of 166.71–168.29 W/m². These values are considered "fairly good" on a pre-existing magnitude categorization assessment.

In Ardabil, highest and lowest wind speeds of 6.58 and 2.09 m/s were found in February and November, respectively. The largest power densities were in autumn, with values of 149.54–150.15 W/m², also earning a rating of "fairly good." Average monthly mean wind speeds in the study period were between 3–5m/s for both cities.

After discussing the raw wind data and analytics, the economic viability of a hypothetical 25 kW wind turbine was considered. To estimate the price of this turbine, initial investment costs were added to yearly maintenance costs. Total costs of turbine implementation, including grid integration, were estimated at $50,000. Several assumptions were made in this estimate, such as an expected life of 25 years and a real interest rate of 22%. Using these estimates, power could be supplied at a monthly varying price of 0.100–0.867 $/kWh in Tabriz, and 0.130–0.477 $/kWh in Arbadil. Current market prices in Iran are 0.18$/kWh for both cities, so wind energy could be competitive for 4–6 months of the year. These calculations exclude potential (and likely) carbon credits that may be granted for renewable energy implementation (as part of a system providing incentives for climate change mitigation).

While these estimates only show cost effectiveness by a narrow margin, they were done in a very conservative manner, as the efficiency of larger turbines (>25kW and/or >10 meters) is much higher. The dense population in Tabriz, along with its oil refinery operations, have created air pollution concerns, therefore, switching to wind energy offers an additional local benefit. This research indicates the likely success of future wind energy projects in these major Iranian cities, and the overall potential for an increased share of renewable energy from Iran.

Wind Turbines Threaten Migratory Bats According to Distribution Models

Wind farms offer a renewable energy source with lower greenhouse gas emissions than traditional fossil fuel sources. Unfortunately, they come with their own set of consequences on local habitats. While previous impact assessments have examined effects on birds, little research has been done in relation to bats. In efforts to infer the effects of wind farms on regional habitat connectivity, Roscioni *et al.* 2014 investigated the effects of a rapidly developing wind energy project in the Molise region of Italy on the bat species *Nyctalus leisleri*. Habitat connectivity has been established as a precursor for biodiversity, a major ecological priority in the public sphere. In this study, bat activity was monitored and potential commuting corridors were identified. These corridors were paired against the current and forecasted wind turbines in the area. By comparing corridors in areas with and without wind turbines, it was observed that bat commuting patterns were altered with the inclusion of the turbines. This paper contributes to a growing body of research surrounding the adverse effects of renewable energy systems on local biodiversity.

Wind turbines can effect bat populations in multiple ways: death from impact, disturbance of migration and/or commuting routes, and habitat loss from construction. While the first and third effects have been detailed in the past, this study explored the little known question of how wind power plants effect commuting routes. The Italian wind farm had 380 currently active turbines with another 195 in planning stages covering approximately 4,460km^2. Within this area, multiple environmental variables were considered alongside the turbines, such as water sources, steep areas, and types of land cover.

The research process consisted of identifying the most suitable areas for *N. leisleri*, identifying corridors between areas, and pairing those corridors with existing and planned wind turbines. Species distribution models (SDMs) and connectivity analysis were used to identify the movements of *N. leisleri*. The SDM was a product of 47 species presence datasets from the area and was used to generate maps of species abundance. The models were carefully calibrated to be most accurate for a small scale dataset. Next, a new UNICOR software program was used to create a "smooth" population density predictions over the whole area. From this density map, probably corridors were modeled. Within these corridors, 150 meter circular barriers were included to simulate the introduction of wind turbines.

The map was shown to have a high degree of predictive capacity. It indicated that 73.4% of the corridors were in the western half of the region, which contained most of the at-risk areas. Overall, 34 planned turbines and 88 existing ones were potentially harmful to *N. leisleri* because of overlap with commuting routes. In this particular region, the corridors are more abundant in the same area as the majority of the turbines. This creates a conflict and illustrates how special attention ought to be paid to the western part of the study area.

This approach demonstrates how researching habitat connectivity can highlight areas of key importance to a species. From there, management suggestions can be made to place turbines in locations with less impact. In the case of the Molise region, it is recommended here that turbines development in the sensitive Western region be limited by avoiding the planned construction of 34 turbines and enacting operational limitations on certain existing turbines.

This study comes with several limitations. First, The long range migratory and adaptive capabilities of *N. leisleri* are unknown. Therefore, it was assumed that these corridors represented daily commuting routes from roosts to foraging sites, but *N. leisleri* is a migratory species, and it is possible that the inclusion of longer range migrations would skew the corridor maps. Limitations aside, this study provides a framework that can be used for impact evaluations on other flying vertebrates.

Wind and Solar GHG Emissions Vary Substantially, But are Lower than Coal or Gas in all Cases

Renewable energy sources, such as wind power generation, are often touted as preferable alternatives to fossil fuels because they produce electricity in an "emissions-free" manner. In actuality, some emissions are created during the production, distribution, and disposal of these technologies, making them not a truly "emissions-free" means of energy production. In order to determine the real relative advantages of various energy sources (in respect to carbon emissions), the full life cycle must be considered. Daniel Nugent and Benjamin Sovacool conducted a literature review of 153 lifecycle studies examining total carbon emissions associated with energy from wind and solar plants and determined estimates of industry averages. Of the 41 studies deemed "best," an average of 34.1 g CO_2/kWh was seen for wind energy and 49.9 g CO_2/kWh for solar. Among these cases, substantial variability was observed, with wind emissions varying between 0.4–364.8 g CO_2/kWh and solar emitting 1–218 g CO_2/kWh.

When conducting the literature review, 153 peer-reviewed papers were initially found that discussed the issue of lifecycle emissions from renewable energy. Of those studies, some were excluded for poor relevance, old dates, unoriginality, and inability to consider all sources of emissions. Studies published before 2003 were not considered. Once narrowed down to a more accurately representative sample, the remaining 41 studies were considered for analysis.

For both solar and wind energy, 71% of the total emissions came from cultivation and fabrication, which consist of multiple stages. This percentage includes resource extraction necessary for the final products and processing the materials —two stages that often involve the use of petroleum. Construction yields another 20% of emissions and the remainder come from operation. Decommissioning actually yields a net *loss* in emissions because the materials are often recycled, especially in the case of wind energy, where "–19.4%" of its emissions arise.

The final average emission values for wind are only part of the story. Most values tend to be much lower, but the averages were brought up by two studies that predicted drastically higher emissions of 138–220 and 364.83 g CO_2/kWh. In the second and highest case, the extra-high emissions came from the storage batteries needed for that system. Several factors were found to reduce emissions in wind energy systems, such as longer lifespan, excluding battery backup, increased height, and building offshore. Storage systems for wind energy are often proposed to help combat intermittent power supply, but the higher emissions from these systems adds complication, especially considering that one main advantage of wind power is its low-carbon status.

Nugent and Sovacool list three main conclusions: first, more scrutiny is needed in the research of emissions from renewable energy systems. Only a small fraction of the existing literature was usable for their study due to various methodological issues. Second, the specific configurations of a wind or solar project can drastically change GHG emissions, and such consequences should be considered. Third, even when including the outlier examples, the average value of 34.1 g CO_2 /kWh for wind energy is drastically lower than coal or natural gas, which have emissions of 960–1050 and 443–611 g CO_2/kWh respectively. So while wind power is by no means "emissions-free," it is still a significantly more climate-friendly option than fossil fuels. It should also be noted that these studies assume energy changes are done in isolation, and that the materials are being produced in a society that still generates power from fossil fuel sources. Emissions would likely be lower if, for example, wind turbines were constructed using energy from other wind turbines.

Conclusions

As humankind continues to search for the most reasonable and effective solutions to climate change, the progress and development of renewable energy projects is of the utmost importance. In order for wind energy to become more widespread, it must be economically viable in more locations. This will require continued research into the most economically efficient energy storage systems, increased weather prediction accuracy, and other policy instruments to incentivize the production of renewable energy. While wind power does contribute some ecological damage of its own, this damage is often minimal. Wind offers a resource that is renewable, available, and relatively eco-friendly. If it can become reliable and economical, it has potential to become a major global energy source.

References Cited

Bandyopadhyay, R., and Patiño-Echeverri, 2014. Alternative energy storage for wind power: coal plants with Amine-based CCS. Energy Procedia 63: 7337–7348.

Chingulpitak, S. and Wongwises, S., 2014. Critical review of the current status of wind energy in Thailand. Renewable and Sustainable Energy Reviews 31, 312–318.

Dan Gao, Dongfang Jiang, Pei Liu, Zheng Li, Sangao Hu, Hong Xu, 2014. An integrated energy storage system based on hydrogen storage: Process configuration and case studies with wind power. Energy, Vol. 66: 332–341.

Fazelpour, F., Soltani, N., Soltani, S., Rosen, M., 2015. Assessment of wind energy potential and economics in the north-western Iranian cities of Tabriz and Arbadil. Renewable and Sustainable Energy Reviews 45, 87–99.

Global Wind Energy Council, 2014. Brussels, Belgium. Report 10.2.2015.

Haessig P., Multon B., Ahmed H. B., Lascaud S., and Bondon P, 2015. Energy storage sizing for wind power: impact of the autocorrelation of day-ahead forecast errors. Wind Energy 18, 43–57.

International Energy Agency, 2014. CO_2 Emissions from Fuel Combustion Highlights 2014.

Konrad Meier, 2014. Hydrogen production with sea water electrolysis using Norwegian offshore wind energy potentials. International Journal of Energy and Environmental Engineering Vol. 5: 1–12.

Mostafaeipour, A. 2010. Productivity and development issues of global wind turbine industry. Renewable Sustainable Energy Review 14, 1048–1056.

Nugent, D. and Sovacool, B.K., 2014. Assessing the lifecycle greenhouse gas emissions from solar PV and wind energy: A critical meta-survey. Energy Policy, Vol. 65: 229–244.

Roscioni, F., Rebelo, H., Russo, D., Carranza, M.L., Di Febbraro, M., Loy, A., 2014. A modelling approach to infer the effects of wind farms on landscape connectivity for bats. Landscape Ecology 29, 891–903.

Sahin, A.D. Progress and recent trends in wind energy. Energy Combust 30, 501–543

Telleria, J.L. 2009. Wind power plants and the conservation of birds and bats in Spain: a geographical assessment. Bio-Diversity Conservation 18, 1781–1791.

Solar Power: Impacts and Sustainability

Jincy Varughese

In the past fifteen years, utility scale solar energy production in the United States has increased by almost 20% (Institute for Energy Research 2015). The Mojave Desert in the Southwestern United States best exemplifies this recent growth. Due to its high levels of solar irradiation and low levels of urbanization, a number of large scale solar energy systems have been built in the Mojave including the Ivanpah Solar Power Facility, the Mojave Solar Project and the Antelope Valley Solar Ranch. In addition to large scale adoption, the domestic use of photovoltaic panels on rooftops has also grown considerably. In the past, attention around solar energy systems has focused on its potential to be an alternative energy source, however given the growing adoption of solar energy, particularly in bio-diverse areas like the Mojave Desert, recent research has concentrated on the sustainability and impacts of solar energy use. A major focus has been on the environmental impacts of utility scale solar energy systems, particularly the high water consumption for cooling and cleaning panels, greenhouse gas emissions, soil erosion, as well as habitat loss and degradation. In regards to urban environments, there has also been research on the effect of rooftop solar panels on local temperatures which are already higher than rural environments due to the "urban heat island" effect. Conversely, another major area of research has focused on the environment's impact on solar energy production and on analyzing how co-locating solar panels on unused open spaces such as in surface waters, airport grass fields, and agave plantations can minimize these environmental impacts while maximizing efficiency. In this chapter, recent research on the environmental impacts of solar energy installations as well as opportunities for colocation will be discussed further.

Environmental Impacts of Utility-Scale Solar Energy

With utility-scale solar energy (USSE) systems growing in number internationally, many have researched the environmental impacts of such systems. Hernandez *et al.* (2014) reviews studies examining the environmental impacts of USSE on biodiversity, water, human health, and potential solutions to mitigate impact.

An indirect effect of USSE is that removal of vegetation as well as the fragmentation of land caused by inserting solar panels and transmission lines can destroy habitats and disrupt the migration of animals, thus disturbing the gene pool. As a result, USSE systems are intentionally sited in areas where biodiversity impact is low. However, in some situations, repatriation and translocation of native plant and animal species are required but these processes generally have a low success rate (<20%), especially for birds, which cannot be easily relocated and are attracted to parts of USSE. Additionally, climate change can alter these habitats even more; predicting these changes and the response of species to climate change involves a high level of uncertainty, thus complicating translocation and repatriation further.

Another disadvantage of USSE is that concentrated solar power systems with wet cooling require copious amounts of water-more water than coal and natural gas consumption combined. This can strain arid environments and can alter the ecosystems in these environments. Utilizing dry cooling which requires 90-95% less water allows a more sustainable option in arid climates.

USSE power plants also pose health risks to humans. Soil erosion from vegetation removal can increase levels of PM2.5 in the air and can spread soil borne pathogens which can potentially contaminate the air as well as water reservoirs. Additionally, photovoltaic cells contain silica dust, cadmium, and arsenic. If these cells are damaged, human exposure to toxics can result. Leaks can also harm human health. USSE power plants involve the use of dust suppressants, cooling liquids, herbicides, and heat transfer fluids-any of which, if leaked into ground water, can have significant public health effects.

However, multiple co-benefit opportunities exist for sustainable USSE development. Utilizing degraded lands avoids some ecological impact issues and co-locating USSE in agricultural areas allows for efficient land use. Additionally, in agricultural areas, livestock can graze around USSE systems, eliminating the need to remove vegetation completely. Floatovoltaics systems are another from of USSE designs which are increasingly implemented globally and conserve water.

It is important to note that these environmental impacts have yet to be weighted against the benefits of utility scale solar energy systems. However, recent research highlights the growing need to address these issues in order to ensure long term sustainability.

Analyzing Bird Use of Photovoltaic Installations at US Airports

Recently, multiple airports in the US have installed photovoltaic (PV) solar energy systems in large areas of undeveloped grasslands commonly found on their grounds. The potential environmental and economic benefits have been well documented but to date there has been no research on the effect such systems will have on birds. DeVault and a team of researchers from the US Department of Agriculture and Mississippi State University begin filling this knowledge gap by comparing bird use of PV arrays to that of nearby airport grasslands.

DeVault *et al.* (2014) observed bird patterns for one year in five locations where PV arrays were near airports. The researchers established bird survey transects at the sites and conducted at least one morning and one afternoon survey each month at each transect. Bird usage and interaction with the airfields and the PV arrays were documented but birds simply flying through the transects were not included in the final analysis. Additionally, in order to analyze whether the PV arrays attracted a larger biomass of birds, a bird hazard index (BHI) was calculated by multiplying the number of each species of bird observed in a transect by its expected mass and then summing up the values for each species observed in the transect.

The researchers found that there was no difference in BHI for the airfields with and without PV arrays, demonstrating that the presence of PV arrays does not alter bird usage, at least when measured by biomass. On the other hand, twice as many birds but fewer species were observed at the PV arrays than at the airfields. This is consistent with other research that large-scale solar development is harmful to wildlife diversity. Additionally, DeVault *et al.* did not find that birds were attracted to the light reflected by the PV panels nor did they observe any collisions with the PV structures. A comparison of seasons shows that BHI was greatest in the summer, reflecting increased bird usage of PV arrays for perching and shade. In fact, most of the observations of PV array transects involved small birds perching on panels. Small birds are not very hazardous to aircraft and small birds that are perching as opposed to flying, are even less hazardous. This suggests that in some locations, implementing PV arrays can actually decrease the risk of bird strikes in comparison to grassy airfields.

Considering Global Dimming and Brightening in Solar Resource Assessments

The planning process for a solar power system requires a solar resource assessment; which involves predicting the annual long-term average solar radiation for the lifetime of the solar system. These values are determined by using long-term data previously collected at other solar power systems. However, the period from the 1950s to 1980s saw a decrease in solar radiation, called global dimming, and since the mid 1980s, the opposite change-global brightening, has occurred. In "Rethinking Solar Resource Assessments in the Context of Global Dimming and Brightening", Müller and his team from Germany analyze the effect of these solar radiation trends on solar resource assessments and explore the effectiveness of other solar radiation prediction methods.

Müller *et al.* (2014) utilize data from 30 German observation stations, collected over more than 40 years. Thermopile pyranometers measured global horizontal irradiance (GHI) and diffuse horizontal irradiance (DIF). From these values, DHI, direct horizontal irradiance was calculated and the effectiveness of utilizing a 10 year, 20 year, and 30 year reference period was analyzed. The researchers found that actual solar radiation during the dimming period was less than would have been predicted using any of the above reference periods and that during the global brightening period, actual solar radiation was greater than would have been predicted. Additionally, during the brightening period, DIF trends are negative while DHI trends are positive. The authors attribute this to a decrease in air pollution and aerosols, which contribute to global brightening but which reduces scattering and absorption that would contribute to diffuse radiation.

Muller *et al.* concluded that although the 10 year reference period underestimates future irradiance, it resulted in the most conservative estimate and thus is the best predictor during periods of global brightening or dimming.

Analyzing Surface Damage of Mirrors Used in Concentrated Solar Power Plants

In Concentrated Solar Power (CSP) plants, 30% of investment costs are attributed to the installation, maintenance, and replacement of mirrors. In order to ensure high yield at a CSP plant, the solar specular reflectance of these mirrors must be maintained at high standards. However, several environmental factors such as temperature, humidity, and windstorms contribute to the degradation of these panels. The latter, is of particular interest to Karim and a team of researchers from Morocco who, over the course of two years, evaluated the effect of windstorms on surface erosion of CSP mirrors.

Karim *et al.* (2014) utilize both natural aging tests and erosion tests in a sandblasting chamber to observe the effects of windstorms on mirror roughness and performance. The natural aging tests occurred on a desert and a seaside site. The sandblasting chamber allowed natural aging conditions to be stimulated but in a controlled environment. Climatic parameters such as wind speed and direction as well as geological parameters such as size distribution, shape, and hardness of sand particles were recorded for these sites. During these different conditions, the specular reflectivity was also measured using a portable reflectometer.

The researchers found that the optical performance of CSP mirrors decreased with increasing wind speed and found that even low wind speeds were associated with ring cracks. The study demonstrated that roughness, impact, and losses in specular reflectivity increased for CSP mirrors with increasing tilt for angles between 20 and 90 degrees. The study also found that particle properties have significant effects on mirror degradation. When controlling for wind velocity and particle size, desert areas with hard sand had mirrors with high roughness parameters whereas mirrors in seaside areas with sharp sand had lower roughness but higher impact parameters. Additionally, course sand was found to contribute to a higher loss in specular reflectivity than fine sand. In conclusion, Karim *et al.* found that wind velocity, mirror angle, particle hardness, and particle sharpness affect the optical performance of CSP mirrors and can contribute to their degradation in natural conditions.

Analyzing Lifecycle GHG Emissions from Solar PV and Wind Energy

Although solar photovoltaic and wind energy are often touted as greenhouse gas emissions-free technologies, the extraction of materials, manufacturing, operation, and decommissioning of these systems release GHGs. Nugent and Sovacool (2014), researchers from the Institute for Energy and the Environment and the Center for Energy Technology respectively, analyzed the level of emissions at each phase of the lifecycle for photovoltaic energy generation systems by conducting a literature search and screening for the most relevant, recent, original, and complete peer reviewed studies.

When looking at each of the phases in its lifecycle, the researchers found that wind energy systems release on average 34.1 g CO_2-

eq/kWh. Furthermore, the majority of GHG emissions lie in the manufacturing stage, which comprises about 71% of wind energy systems' emissions. The construction and operation stages accounts for about 24% of lifecycle emissions. Interestingly, because decommissioning of wind energy systems involves recycling and reusing materials, this phase offsets 19.1% of wind energy system emissions.

PV panels on the other hand generate an average of 49.9 g CO_2-eq/kWh, which is higher than that of wind systems. The manufacturing, construction, and operation phases compose 71%, 19%, and 13% of lifecycle emissions with the decommissioning stage offsetting 3.3%.

The main contributors to high levels of GHG emissions during the manufacturing stage were resource inputs, technology, and location. The researchers found that PV panels release a range of emissions depending on the type of material used-with mono-Si panels having the highest emissions and panels made of CdSe QDPV having the lowest emissions. Additionally, the researchers found that location of manufacturing also contributes to variations in emissions. For example, manufacturing wind turbines in Germany would release fewer emissions than manufacturing them in China because China still relies much more heavily on coal for electricity.

Ultimately, the researchers concluded that more rigorous studies on the lifecycle emissions of solar and wind energy must be done, particularly related to optional energy storage, which improves efficiency but at the cost of greater GHG emissions. However, this study identified areas such as manufacturing, location, and materials, where improvements can result in GHG emissions reduction for wind and solar PV energy systems

Solar Panels Reduce Urban Heat Island

It is well known that on a global scale, increasing number of solar energy production will reduce greenhouse gas emissions and will thus reduce global warming. Lesser known, however, are the local climatic impacts of increasing solar panels, especially in cities. Cities tend to be warmer than countryside areas, creating what is known as the Urban Heat Island (UHI). The effects of the UHI are expected to increase with climate change, causing city planners to strategize the use of certain building materials over others, but little research has been done on the role of solar panels in contributing to changes in city temperatures. A team of researchers from France has addressed this knowledge gap by using models to estimate the impact of solar panels on the UHI in cities. They found that the presence of solar panels in cities actually helps mitigate the impact of the UHI.

Masson *et al.* (2014) utilized the Town Energy Balance scheme to assess the energy production from solar panels and then accounted for characteristics such as albedo, emissivity, and proportion of roof covered by solar panels. A simulation was done for the Paris metropolitan area and also on the countryside, for a reference. After running the simulation for summer months, the researchers found that the presence of solar panels decrease air temperatures in Paris by more than 0.2 K during the day and by 0.3 K at night. The higher cooling rate at night occurs because solar panels intercept radiation and reduce heat stored by buildings. Additionally, the thinner urban boundary layer at night also contributes to the greater cooling effect. Because of this cooling effect, the presence of solar panels increased the amount of energy needed for heating in the winter months by 3% and decreased the energy needed for air conditioning by 12%. In summary, Masson *et al.* found that with urban temperatures expected to rise, solar panels can reduce the urban heat island and decrease the demand for air conditioning.

Analysis of Floating Offshore Solar Power Plants

In Europe, large flat spaces of land with direct normal irradiance (DNI) levels high enough to support solar utilities are uncommon. Diendorfer *et al.*(2014), a team of researchers in Vienna, saw untapped potential in the Mediterranean Sea and began researching the feasibility of offshore solar power plants. Aside from the availability of open spaces, building solar power plants offshore has two main advantages. First, a system that revolves on a vertical axis can be easily implemented and is efficient at sun-tracking. Second, unlimited water is readily available for the cooling processes required at solar thermal power plants. In this paper, the use of both Parabolic Trough Collectors (PTC) and Pneumatic Pre-Stressed Solar Concentrators (PPC) are considered. The study evaluates the performance of a preliminary floating solar power plant using a model that accounts for platform motion, sea state, and solar irradiance.

As the motion of the waves can change the angles of the panels, thus affecting the efficiency of solar power plants, sea state along with the irradiance levels must be included as variables in the model. Hourly direct solar irradiance values for the Mediterranean Sea were calculated using data on global horizontal irradiance from 1991 to 1993. Sea state variables which include wave height, mean wave period, and wave direction were also incorporated into the model as was platform motion.

The study found that wave motion and direction has little effect on the efficiency of the power plant during normal operating conditions. Diendorfer *et al.* did find that efficiency falls in the presence of choppy waves but the performance loss in comparison to land systems is small and also results from a reduction in DNI during stormy weather as well. The study also found that using PPC in offshore solar power plants was more effective than PTC in ensuring alignment, lower in cost, and more resistant to salt water. Although offshore solar power plant development is still in its early stages, Diendorfer et al. conclude that offshore plants are an efficient and feasible means of increasing Europe's energy supply.

Tradeoffs and Synergies in Co-locating Biofuel Production and Solar Panels in Deserts

The development of solar energy systems in desert environments has grown in recent years along with levels of biofuel production. Although utility scale photovoltaic energy systems in deserts have low greenhouse gas emissions, they require large amounts of water in suppressing dust from disturbed soil during construction and cleaning panels during operation. The majority of this water runs off into the soil. Co-locating biofuel feedstocks such as agave, which can be fermented into ethanol and can be sustained on very little water, can be an effective use of this runoff water while utilizing land in a way that is not competitive with traditional agriculture. Ravi and a team of researchers from Stanford University have done a life cycle analysis of the advantages and disadvantages of a hypothetical co-located solar-agave system.

To conduct lifecycle analyses, the researchers compiled data from existing large scale solar installations in the Southwest and since the use of agave in biofuels is not well established, the researchers utilized data from the tequila industry in Mexico and the Brazilian sugarcane industry.

The analysis of the hypothetical colocation of solar PV and agave revealed higher economic returns per cubic meter than the use of either system alone. The researchers attribute this to efficient water use and the mitigation of dust generation due to the ability of agave plants and vegetation to prevent soil erosion. Through their baseline yield scenario, the researchers found that agave can be sustained using only the water used for cleaning solar panels; although the energy output of this yield is much lower than that of PV systems, the economic advantages are high as agave is easily transportable, cultivation is land use efficient, and requires low amounts of water. The researchers note that some utility scale solar energy systems in the United States could integrate agave cultivation sites as well through some design modification and regular agricultural upkeep. If this were to be done on all existing and future PV sites, colocation could result in the production of 1-2 billion gallons of ethanol, all while efficiently utilizing underused land and scarce water resources.

Conclusions

This research points to the need for greater consideration of environmental impacts in the planning and implementation of solar energy production systems. In particular, the research illustrates that greater efficiency in water use, lower lifecycle GHG emissions, and minimal disturbances to vegetation and animal habitats are priority areas for sustainable solar development. Current research also illustrates that there are several methods such as placing solar panels in areas with low wind speeds or co-locating solar installations with agave plants, which can not only help reduce these impacts, but can potentially be implemented in the near future. However, the paucity of studies in this field highlights the "newness" of this focus area as well as solar energy industry and points to the need for more exhaustive research in order to better understand the environmental impacts of solar energy and the best methods to mitigate them.

References Cited

Devault, T., Seamans, T., Schmidt, J., Belant, J., Blackwell, B., Mooers, N., Tyson, L., Van Pelt, L."Bird Use of Solar Photovoltaic Installations at US Airports: Implications for Aviation Safety." *Landscape and Urban Planning*: 122-28.

Diendorfer, C., M. Haider, and M. Lauermann. 2014. Performance Analysis of Offshore Solar Power Plants. *Energy Procedia* 49, 2462-471. [GSSS: performance offshore solar power plants] http://www.sciencedirect.com/science/article/pii/S1876610214007152.

Hernandez, R., Easter, S., Murphy-Mariscal, M., Maestre, F., Tavassoli, M., Allen, E., Barrows, C., Belnap, J., Ochoa-Hueso, R., Ravi, S., Allen, M. "Environmental Impacts of Utility-scale Solar Energy." *Renewable and Sustainable Energy Reviews*: 766-79.

Karim, M., S. Naamane, C. Delord, and A. Bennouna. "Study of the Surface Damage of Glass Reflectors Used in Concentrated Solar Power Plants." SolarPACES 2014.

Masson, V., Bonhomme, M., Salagnac, J., Briottet, X., Lemonsu, A. 2014. Solar panels reduce both global warming and urban heat island. Frontiers in Environmental Science 2, 1-10.

Müller, B., Wild, M., Driesse, A., Behrens, K. "Rethinking Solar Resource Assessments in the Context of Global Dimming and Brightening." *Solar Energy*: 272-82.

Nugent, D., Sovacool, B., 2014. Assessing the Lifecycle Greenhouse Gas Emissions from Solar PV and Wind Energy: A Critical Meta-survey. Energy Policy, 229-244.

Sujith, R., Lobell, D., Field, C., Tradeoffs and Synergies between Biofuel Production and Large Solar Infrastructure in Deserts. Environmental Science and Technology 48. 3021-3030.

"Topics: Solar". Institute for Energy Research.

Hydraulic Fracturing in the United States

Alex Frumkin

Hydraulic fracturing, or fracking, is a technology that has existed since the late 1940s but has become increasingly popular in the past two decades. Fracking is the process of injecting millions of gallons of water, sand, and chemicals deep into the ground to fracture rock beds and allow for gas and oil to flow back up the well. Fracking allows gas and oil companies to access previously inaccessible shale beds or shale beds that were previously not economical to tap into. Technological advances in the fracking process have made fracking significantly less expensive and a technology that is used at most new wells across the United States.

Hydraulic fracturing is not currently federally regulated in the US, so each state is responsible for creating its own regulations around fracking. This has led to a huge variance in the stringency and the complexity of fracking regulations, and made it impossible for companies to anticipate how different states will choose to regulate fracking.

Due to the recent growth of fracking, and the fact that it is not federally regulated, there is still much that is unknown about its side effects, and there are some major adverse public health effects that could be associated with the process. While the link is not perfectly clear yet, there has been a significant amount of research showing fracking leading to increased respiratory health problems for residents near fracking wells. Another major issue that is explored in this chapter is the possibility that hydraulic fracturing leads to water contamination, which can negatively affect human health.

One of the reasons that hydraulic fracturing is considered favorable is that natural gas has a lower carbon footprint than other fossil fuel energy options. Many fracking proponents suggest that this is a good intermediate step towards a cleaner carbon future. However, research discussed in this chapter uses predictive models to show that natural gas will not only be replacing coal but also cleaner renewable energy sources, casting doubt on this claim.

This chapter gives only a small glimpse into some of the research that is being done in this rapidly expanding field of study.

Methane Contamination of Drinking Water Accompanying Gas-Well Drilling and Hydraulic Fracturing

Directional drilling and hydraulic-fracturing technologies are dramatically increasing natural-gas extraction across the United States. Hydraulic fracturing remains largely unregulated at the Federal level regardless of the growing concerns about contamination of drinking water. However, the potential contamination risks in shallow drinking-water systems are still not fully understood, and a topic of study for many scientists. There are four main reasons why scientists and public health officials are concerned about methane contamination in the ground water: that the chemicals use in fracturing fluid can leak into the ground water, that the water can become explosive if methane levels are high enough, that the methane could be released into the environment, and that the untested and unregulated shallow ground water in rural areas near drilling sites could be ingested during household or agricultural use. Scientists have continued to study whether water wells are being contaminated in any of these ways by hydraulic fracturing and drilling.

This paper analyzed the aquifers overlying the Marcellus and formations of northeastern Pennsylvania and Upstate New York and whether there is evidence for methane contamination of drinking water associated with shale-gas extraction. Gibson *et al.* (2011) studied the presence of dissolved salts, water isotopes, and isotopes of dissolved carbon, boron, and radium in ground water from 68 private water wells ranging in depth from 36- to 190-m deep. In addition, 60 of these 68 wells were also analyzed for dissolved-gas concentrations of methane. Methane concentrations were detected in 51 of 60 drinking-water wells, but concentrations were substantially higher closer to natural-gas wells. Methane concentrations were 17-times higher on average in shallow wells from active drilling and extraction areas (areas with multiple drilling and extraction areas within 1 km of each other) than in wells from non-active areas. The average concentrations found near the drilling sites fell within the defined action level for hazard mitigation.

In order to understand the origin of the methane found in the water it must be determined whether the methane is coming from shallow biogenic or deeper thermogenic contamination. The presence of ethane and propane indicates that the methane is a result of deeper thermogenic contamination. Ethane was found in only 3 of the 35 water wells from non-active drilling sites, whereas it was detected in 21 of 26 water wells in active drilling sites. This indicates that the high methane concentrations observed are most likely due to thermogenic gas. There are multiple explanations for why methane concentrations may be increased near gas wells, but these results suggest the need for long-term sampling and monitoring of the hydraulic fracturing and drilling industry and private water wells. In addition, these results support the need for regulation of the hydraulic fracturing at the Federal level.

The Risk that Hydraulic Fracturing Poses to Water Sources

There has been a rapid increase in shale gas development in the United States due to the increase in use of hydraulic fracturing to access these shale beds. The rise of hydraulic fracturing has lead to intense public debates about the potential environmental and human health effects from hydraulic fracturing. Vengosh *at el.* (2014) identifies four potential areas of risks for water resources from hydraulic fracturing: contamination of shallow aquifers due to stray gas contamination, contamination of surface water and shallow groundwater from spills, leaks, and/or the disposal of inadequately treated shale gas wastewater, accumulation of toxic and radioactive elements in soil near disposal or spill sides, and the over extraction of water resources that could induce water shortages. To be able to fully understand the water contamination risks associated with hydraulic fracturing there needs to be an in depth investigation of the hydrology, hydrogeology, water chemistry, and isotopic tracers for identifying what the cause of the water contamination is.

The review of the literature identified four possible risks to water resources. Hydraulic fracturing could lead to stray gas leaking into shallow aquifers due to poorly constructed or failing gas wells. This process could happen over a long time so any current evidence of stray gas contamination could be a clue for future water degradation. Water resources could also be contaminated in areas where hydraulic fracturing is occurring because of spills, leaks, or disposal of hydraulic fracturing fluids and/or inadequately treated wastewater. Both of these are full of chemicals that can be highly damaging for both human and environmental health. There's also the possibility of metal and radioactive element accumulation in water-ways due to disposal or spill sites. This is problematic because this accumulation can release toxic elements and radiation into the environment. The last identified risk is the overexploitation of limited or diminished water resources due to huge withdrawals of valuable fresh water that is necessary for hydraulic fracturing to be done.

There is still debate around whether hydraulic fracturing does pose these risks to water resources, and if the risks are there, how severe the risks really are. In order to put an end to this debate, it's necessary to develop a data for baseline water chemistry in aquifers before hydraulic fracturing begins. It is imperative that a comprehensive investigation is completed on the hydrology, hydrogeology, water chemistry, and isotopic tracers to identify whether water contamination has occurred near hydraulic fracturing sites. In order to fully put a rest to this debate, these investigations need to happen at shale gas developments across the country, not just at a few geographic locations.

Time Frames on Upward Migration of Hydraulic Fracturing Fluid and Brine

As hydraulic fracturing becomes more popular in the United States, there has been growing concern about possible upward migration of hydraulic fluid and brine into U.S. drinking waters. This paper written by Flewelling and Sharma looks at the constraints on this upward migration from black shale to shallow water aquifers due to hydraulic fracturing and the time frames in which this upward migration could occur. Ultimately, the authors show that there are few hydraulic fracturing sites where there is both an upward gradient and permeability is low, and when those rare cases do exist the mean travel times are long (often $>10^6$).

Flewelling and Sharma's study focused on permeabilities, head gradients, and the relationship between the two. By studying theses factors they illustrate that in situations where upward head gradients exists, permeability is low, and the chances of vertical fluxes are low. In addition to the low vertical fluxes the timescales for transport are long. This is because upward migration of hydraulic fracturing fluids would require both high bedrock permeability and upward head gradients, however, the authors point out, that these two conditions are mutually exclusive. This demonstrates that widespread upward migration of hydraulic fracturing fluid and brine is not physically plausible.

While there are some mechanisms that could cause the upward migration, those mechanisms are not commonly present in black shale hydraulic fracturing sites, and when they are the flow rates are low and flow paths are long. Water that penetrates deep enough to drive brine migration would typically take tens to hundreds of years before re-entering at the surface and topographically driven flow of brine to the surface is associated with travel times of at least a few million years.

This paper illustrates that the recent concern about rapid upward migration due to hydraulic fracturing is unfounded because the analysis of the black shale indicate that where upward flow occurs, both permeability and flow rates are low which leads to long timescales for transport.

Is there a relationship between proximity to natural gas wells and health?

There has been little research about the public health impacts of living near unconventional natural gas extraction activities. Rabinowitz *et al. (2015)* aimed to assess a possible relationship by generating a health symptom survey of 492 people in households with ground-fed wells in an area of active natural gas drilling. The survey looked at the household's proximity to gas wells and then the prevalence and frequency of reported dermal, respiratory, gastrointestinal, cardiovascular, and neurological symptoms. The study found that individuals who lived within 1 km of a gas well were twice as likely to experience upper respiratory systems than individuals in households more than 1 km away. No relationship found between well proximity and any of the other possible health conditions that this survey covered.

The increase in unconventional methods of natural gas extraction has lead to concern over the possible adverse public health effects related to this drilling. There are currently very few peer-reviewed studies of the health effects that could be related to the process of natural gas extraction, potential water exposures, and potential air exposures that occur because of hydraulic fracturing. There is concern that these different exposures could happen through contaminants in the water or in the air, and this report was an analysis of a cross-sectional, random-sample survey of the health of residents who had ground-fed water wells near natural gas extraction activities.

The study focused on Washington County, Pennsylvania near where natural gas extraction from the Marcellus Shale is occurring. This area was classified as active because there were 624 active natural gas wells in this county, and 95% of them were horizontally drilled.

The health assessment was created with questions that were drawn from publicly available surveys and on reported health symptoms. The randomly selected households were visited to ensure that they did have a ground-fed water well and that the homes were occupied. The report classified each household surveyed by the household's distance from a gas well: <1 km, 1-2 km, >2 km.

The results from this study showed that sixty-six percent of these households used their ground-fed water for drinking water and 84% used it for other activities such as bathing. The report showed that the average number of reported symptoms per person in households less than 1 km from a gas well was significantly greater than for those living more than 2 km away from gas wells. Individuals in households less than 1 km from natural gas wells were more likely to have skin conditions and upper respiratory symptoms in the past year. The other symptoms that were studied did not show a significant relationship between the proximity to gas well and the frequency of the health symptom. Although a significant relationship was not found between all of the health concerns and proximity to a gas well, this study affirms the need for further research into the health effects of natural gas extraction activities.

Distance: A Critical Aspect for Impact of Hydraulic Fracturing

There has been little research about the public health impacts of living near unconventional natural gas extraction activities. Rabinowitz *et al. (2015)* aimed to assess a possible relationship by generating a health symptom survey of 492 people in households with ground-fed wells in an area of active natural gas drilling. The survey looked at the household's proximity to gas wells and then the prevalence and frequency of reported dermal, respiratory, gastrointestinal, cardiovascular, and neurological symptoms. The study found that individuals who lived within 1 km of a gas well were twice as likely to experience upper respiratory systems than individuals in households more than 1 km away. No relationship found between well proximity and any of the other possible health conditions that this survey covered.

The increase in unconventional methods of natural gas extraction has lead to concern over the possible adverse public health effects related to this drilling. There are currently very few peer-reviewed studies of the health effects that could be related to the process of natural gas extraction, potential water exposures, and potential air exposures that occur because of hydraulic fracturing. There is concern that these different exposures could happen through contaminants in the water or in the air, and this report was an analysis of a cross-sectional, random-sample survey of the health of residents who had ground-fed water wells near natural gas extraction activities.

The study focused on Washington County, Pennsylvania near where natural gas extraction from the Marcellus Shale is occurring. This area was classified as active because there were 624 active natural gas wells in this county, and 95% of them were horizontally drilled. The health assessment was created with questions that were drawn from publicly available surveys and on reported health symptoms. The randomly selected households were visited to ensure that they did have a ground-fed water well and that the homes were occupied. The report classified each household surveyed by the household's distance from a gas well: <1 km, 1-2 km, >2 km.

The results from this study showed that sixty-six percent of these households used their ground-fed water for drinking water and 84% used it for other activities such as bathing. The report showed that the average number of reported symptoms per person in households less than 1 km from a gas well was significantly greater than for those living more than 2 km away from gas wells. Individuals in households less than 1 km from natural gas wells were more likely to have skin conditions and upper respiratory symptoms in the past year. The other symptoms that were studied did not show a significant relationship between the proximity to gas well and the frequency of the health symptom. Although a significant relationship was not found between all of the health concerns and proximity to a gas well, this study affirms the need for further research into the health effects of natural gas extraction activities.

Public Perceptions of Hydraulic Fracturing

Hydraulic fracturing is considered controversial for many reasons, including the possible negative environmental impacts, the possible economic benefits of development, and reduction of reliance on foreign oil. Previous national opinion polls have indicated that a sizable minority of the population lack familiarity with this largely unregulated field. Boudet *et al.* (2014) studied different socio-demographic indicators will predict support of or opposition to hydraulic fracturing.

This study on which socio-demographic indicators will predict support of or opposition to hydraulic fracturing was completed in September of 2012, with a total of 1062 respondents. The conclusions that Boudet *et al.* came to had a margin of error of 3% at the 95% confidence level. Identifying as a woman or having egalitarian worldviews was negatively associated with support for hydraulic fracturing. Individuals who identified as politically conservative was a positive indicator of support. These results were what the researchers were expecting, however formal education which was believed to negatively predict support, ultimately ended up being a positive predictor for support. Similarly, they found that household income, race, and individualist worldview were all not predictive of support, in contrast to the authors' original hypothesizes.

These results indicate that the American populace is largely not knowledgeable of and undecided about hydraulic fracturing. The Americans who are in opposition to hydraulic fracturing are more likely to be familiar with hydraulic fracturing and include: women, egalitarians, people who read newspapers more than once a week and often reference environmental concerns for their position. On the other hand, supporters of hydraulic fracturing tend to be older, have higher education, be politically conservative, watch TV for their news instead of reading newspapers, and their support stems from the possible economic developments. These socio-demographics are critical to understand because history has demonstrated that public attitudes place a large role in how unconventional oil/gas reserves are developed. This research illustrates the need to pursue a wide–ranging and inclusive public dialog about the realities of hydraulic fracturing in America.

What can CCS Learn from Hydraulic Fracturing Acceptance?

Carbon capture and storage (CCS) faces potential obstacles when it comes to the development and deployment of the technology. Many of these challenges are strikingly similar to those faced by proponents of hydraulic fracturing, especially the challenge of social acceptance of this technology. Due to these similarities, Wolff *et al.* 2014 uses hydraulic fracturing as a comparison to identify potential strategies for future carbon capture and storage efforts. When using hydraulic fracturing industry as a comparison the authors consider not only the act of fracturing, but also the process of obtaining mineral rights and the waste removal process. This comparison is achieved by completing statistical analysis on the relationship between state demographics and the stringency of state regulations of the hydraulic fracturing industry. Ultimately, the authors find that states that are familiar with the oil and gas industry have less variable regulation of hydraulic fracturing. In addition, they recognize a disconnect between the regulations of hydraulic fracturing at the state level and at the local level. This tension suggests that carbon storage proponents should focus on local engagement not just on state level.

These two technologies face similar issues around social acceptance since they have many of the same operations, risk, and narratives for the potential benefits and drawbacks. Similar to hydraulic fracturing, social acceptance is critical for CCS to be successful. This includes the acceptance of homeowners, communities, firms that carry out carbon storage, investor, regulators, politicians, voters, various interests group, and the media. Each of these groups has the power to determine the development of CCS through supporting or protesting certain policies or regulations. Since the technology for CCS is not yet mature it is necessary to predict future trends in social acceptance for CCS on the hydraulic fracturing industry, and the way that social acceptance has effected the technology deployment and development.

The statistical analysis performed on a state's familiarity with oil and gas and the state's hydraulic fracturing regulation demonstrated that states with higher familiarity with oil and gas development had a similar level of regulation stringency for hydraulic fracturing as one another. On the other hand, states that were not familiar with oil and gas extraction were more unpredictable on the type of regulation the state would have for hydraulic fracturing. This suggests that carbon storage firms should focus first on opening in states familiar with oil and gas extraction. This is because they are more likely to have predictable regulations than states not familiar with oil and gas extraction.

The authors also performed a case study on the differences between New York and Pennsylvania's regulation of hydraulic fracturing. These two states along with being neighbors are also both over the Marcellus Shale, one of the largest shales in the United States. Although Pennsylvania has greatly expanded their use of hydraulic fracturing, New York placed a moratorium on hydraulic fracturing within the state year after year. While New York had Pennsylvania vary significantly in the hydraulic fracturing allowed within their borders, both states have used zoning laws and local bans to control hydraulic fracturing at the local level. Both states indicate that municipalities have increasing regulatory control over hydraulic fracturing. This should be an indication to the carbon storage industry that municipal ordinances may be as important as state regulations, and that storage firms should be ready to engage stakeholders at a municipal level.

The case study of New York and Pennsylvania also illustrates how political pressure and the national spotlight compelled New York's governor to continue to renew the moratorium on hydraulic fracturing again and again. Carbon storage firms will need to move quickly when it comes to engaging those in power to keep the issue from rising to the nation's attention.

Will Increased Natural Gas Usage Decrease the Effects of Climate Change?

The improvement of hydraulic fracturing technologies in the last decade has allowed access to previously uneconomic shale gas resources across North America. Natural gas production is often touted as a way to cut carbon emissions to slow down climate change because gas-fired power plans emit roughly half as much CO_2 per unit of energy produced as coal-fired plants. There are some assessments that have been completed, though, that argue that natural gas lifecycle emissions are actually higher than those of coal because of emissions from shale gas production. In line with this latter idea, Mcjeon *et al.* (2014), show that market-driven increases in unconventional natural gas production does not discernibly reduce the trajectory of greenhouse gas emissions or climate forcing.

The five models used to come to this result integrate energy, economy and climate systems to provide a consistent framework. The models share common natural gas supply curve assumptions but are different in their architecture, geospatial resolution, socioeconomic assumptions, and technology projections. All of the models showed that abundant gas supply leads to additional gas and gas-fired electricity consumption. The five models also consistently showed that additional supply of natural gas in the energy market does not discernibly reduce fossil fuel CO_2 emissions, and the authors identified two reasons why. The first is that the gas will substitute for all other primary fuels—such as nuclear and renewables—not just coal. This means that it is not just a substitution between emissions factors of gas and coal. The second is that lower natural gas prices will increase economic activity and reduce any incentive to invest in energy-saving technologies, which ultimately will lead to an expansion of the total energy system, and of fossil fuel use.

These models demonstrate what would happen if the market forces were allowed to work themselves out, however the results would be different if there were policies or laws put into place that limit natural gas's ability to substitute for low-carbon energy sources. This study highlights the need for future work on the effectiveness of policies that could be implemented to improve the chances of reducing greenhouse gas emissions.

Conclusions

Hydraulic fracturing is a process that has become increasingly important to access natural gas and petroleum in the United States, but there is still much that is unknown about the process and its consequences. Hydraulic fracturing is becoming increasingly more attractive as it makes oil and fossil fuel resources that were previously out of reach feasible. While fracking may have lower carbon footprints than some of the other extraction methods, there are serious threats to public health that are not currently understood. The threat of water contamination due to mistakes in the fracking process or methane traveling to water sources is not currently understood enough to properly mitigate these risks. Hydraulic fracturing is a booming industry in the United States but one that requires significantly more research before there can be confidence that this process can be done safely.

References Cited

Boudet, H., Clarke, C., Bugden, D., Maibach, E., Roser-Renouf, C., Leiserowitz, A., 2014. "Fracking" controversy and communication: Using national survey data to understand public perceptions of hydraulic fracturing. Energy Policy 65, 57-67.

Flewelling, S.A., Sharma, M., 2014. Constraints on upward migration of hydraulic fracturing fluid and brine. Groundwater 52, 9-19.

McJeon, H., Edmonds, J., Bauer, N., Clarke, L., Fisher, B., Flannery, B.P., Hilaire, J., Krey, V., Marangoni, G., Mi, R., 2014. Limited impact on decadal-scale climate change from increased use of natural gas. Nature 514, 482-485.

McJeon, H., Edmonds, J., Bauer, N., *et al.,* 2014. The Extractive Industries and Society, 124-126.

Osborn, S.G., Vengosh, A., Warner, N.R., Jackson, R.B., 2011. Methane contamination of drinking water accompanying gas-well drilling and hydraulic fracturing. proceedings of the National Academy of Sciences 108, 8172-8176.

Quinmin, M., Steve, A., 2014. Distance: A critical aspect for environmental impact assessment of hydraulic fracking. The Extractive Industries and Society 1, 124–126

Rabinowitz, P.M., Slizovskiy, I.B., Lamers, V., Trufan, S.J., Holford, T.R., Dziura, J.D., Peduzzi, P.N., Kane, M.J., Reif, J.S., Weiss, T.R., 2015. Proximity to natural gas wells and reported health status: Results of a household survey in Washington County, Pennsylvania. Environmental health perspectives 123, 21-26.

Vengosh, A., Jackson, R.B., Warner, N., Darrah, T.H., Kondash, A., 2014. A critical review of the risks to water resources from unconventional shale gas development and hydraulic fracturing in the United States. Environmental Science & Technology 48, 8334-8348.

Wolff, J., Herzog, H., 2014. What lessons can hydraulic fracturing teach CCS about social acceptance? Energy Procedia 63, 7024-7042.

Why Fracking Works—and Sometimes Doesn't

J. Emil Morhardt

We hear a great deal about the economic benefits of hydraulic fracturing, and even more about its potential liabilities, but seldom very much about exactly how fracking works. A fascinating paper published by the American Society of Mechanical Engineers (Bazant *et al.* 2014) combines an extremely clear explanation of the process in non-technical language with a detailed mathematical analysis of the mechanics involved (a combination uncommon in engineering papers). The question at hand is why, with pipes just three-inches in diameter, spaced half a kilometer apart, it is possible to get so much gas out of shale beds. The first thing to know is that even this technology gets only about 5–15% of the gas embedded in the shale, so it's likely they'll be going back for more as the technology improves. They know about this percentage because of how much gas they can extract from the rock samples they get out of the well cores.

The reason fracking is useful in the first place is that the gas, in the form of solid kerogen, is trapped in nanovoids, and the natural cracks that would let it migrate to a borehole are either squeezed shut by the weight of the three kilometers of rock above them, or filled up with calcite or other minerals. The trick is to open up the existing cracks, form new ones, then keep them open so gas can flow out. Through a lot of detailed engineering calculations the authors determine that the way to open up the most cracks (and in the process to keep the most fracking fluid stuck in the shale, rather than returning to the surface where it needs to be treated or re-injected) is to gradually increase the hydraulic pressure to a point where it gradually opens up vertical cracks, while the sand they also inject gradually fills an ever-widening array of them. The acid included in the fracking fluid apparently breaks up the rough edges (asperities) of the cracks into pieces small enough to act like the sand, increasing the amount of crack propping. Since the injection pressure of the fracking fluid seldom exceeds the pressure exerted by the overburden, horizontal cracks in the natural bedding plane are equally seldom opened up. The whole procedure at this point in its evolution is dependent on vertical cracks.

Methane, which makes up most of the content of natural gas, does not dissolve in water (nor in fracking fluid, which is 99% water) so it migrates through the cracks and into the well casing in the form of gas bubbles, which, when they reach the vertical borehole are aided in coming to the surface because of their buoyancy.

The graphs included in the paper show that the highest rate of gas flow occurs soon after a well is drilled, with exponentially decreasing flow over the four- or five-year life of the well.

So, in summary, fracking usually occurs in shale layers 20 to 150 meters thick, lying on the order of three kilometers below ground—a long way below any aquifers that might be susceptible to contamination, but of course the well shaft must pass through these aquifers. From each vertical borehole, several horizontal ones several kilometers long are drilled about 500 meters apart. These horizontal holes are lined with three-inch diameter steel pipe that is then perforated explosively at five to eight locations along their lengths. Several million gallons of water with a little sand, acid, and other chemicals is injected under great pressure. (Although this sounds like a lot of water it corresponds to only 1-2 millimeters of rain falling over the well field.) The injection initially opens up natural vertical cracks in the shale, which are typically 15 to 50 centimeters apart. Continued pressure increases the array of vertical cracks, ideally about 10 centimeters apart. The methane forms bubbles and flows along the pressure gradient toward the horizontal pipe, then through the pipe to the vertical borehole, and up to the surface to be captured.

If you're not familiar with materials science and engineering technical writing, I'd suggest looking at the full paper (link in the References

Cited section.) It opens a window into the thought processes of engineers that most of us never encounter.

Why Fracking Might Not Work for as Long an We Would Like

The December 4, 2014 issue of the scientific journal *Nature* takes the position that the current abundance of natural gas in the US derived from horizontal drilling and hydraulic fracturing may be a much shorter-term phenomenon than most analysts have thought. In both an editorial and an opinion piece (not however in a scientific paper) the journal takes issue with the US Energy Administration's (USEA) assessment that natural gas production in the US will continue to grow for a quarter century, at least. *Nature* relies on the opinions of a team of researchers at the University of Texas, and cites a paper (Patzek, 2012) by members of the team which now consists of a dozen geoscientists, petroleum engineers, and economists. That paper examines extraction data from 2,057 such wells in the oldest US shale play, the Barnett Shale in Texas, and concludes that they started to decline at an exponential rate in ten years or less, and goes on to predict the total amount of gas that will be produced by their overall sample of 8,294 wells; 10–20 trillion standard cubic feet over the next 50 years.

The team producing the paper talked to Nature's Mason Inman (2014) about current work that is only beginning to appear at conference presentations and in scientific journals, but which suggests that the major current US shale gas operations would peak in 2020, and decline from then on, producing only half as much gas by 2030 as predicted by the USEA, even under its most conservative scenarios. This discrepancy may be attributable to the more detailed look at producing wells taken by the Texas team, and the USEA is likely to do similar analyses itself. The bottom line, though, seems to be an increase in uncertainty (not a big surprise to anyone who has done scientific research—the more closely one looks at a problem, the more complicated it becomes, usually.) So one might conclude that the research should serve to temper the euphoria on the part of those profiting from the lower gas prices, as well as the frustration felt by producers as they watch gas prices fall, and by environmentalists who fear that the low prices will stifle attempts to replace fossil fuels with renewables.

Tracking Fracking Fluid with Molecular Tracers

Stephanie Kurose, a law student at the American University in Washington DC, calls our attention to both the concept of, and two startups trying to push, micro-tracers which could be injected into fracking fluid so that if it escapes, we know whodunit. The idea is simple, if not yet operational; create some long-lived non-toxic chemical compound with enough potential variation that a different version could be mixed in with the fracking fluid for each individual well. The arguments for it, espoused by Kurose, are equally simple; drilling companies would know if they had a problem with leakage and could change their technology, falsely-accused drilling companies could exonerate themselves, and the public should feel much less angst about fracking if evidence of leaked fracking fluid fails to materialize (or vice versa.) It might be that drilling companies would resist in order to avoid any conclusive evidence that their wells have leaked, but so far no one knows because suitable tracers have yet to be deployed. The two startups giving it a shot are BaseTrace and FracEnsure.

BaseTrace uses genetic engineering technology to produce strands of resilient DNA which can be readily customized into a nearly infinite number of variations which could be mixed with all sorts of industrial fluids, including fracking fluid. Genetic engineering technology makes it equally simple to read the genetic code in these [relatively short by biological standards] strands of DNA.

FracEnsure uses nanoparticles with a paramagnetic coding that is somehow individually coded in batches, but the company's website does not explain the technology further so we will have to wait.

Instead of Flaring Natural Gas at Fracked Oil Wells, Use it to Treat Fracking Fluid

Seems like a good idea. Yael Rebecca Glazer suggested it in a 2014 Masters Thesis in Engineering at the University of Texas at Austin. A major issue with fracking is that sometimes a lot of the fracking fluid that was pumped down the well to create the fractures comes back up, sometimes along with additional "produced" water, sometimes twice as much as was pumped down in the first place. On top of that, it is often so contaminated that it exceeds the capabilities of industrial treatment facilities, so it gets trucked to a nearby injection well and is reinserted. But injection wells are not always handy, and anyway, the water itself would be valuable if it weren't so polluted. Meanwhile, although a fracked well might producing mainly oil, there is also often a fair amount of natural gas produced; but if there isn't enough gas to make it economical to capture it and sell it, it is commonly flared—burned right there at the wellhead. This converts the natural gas to CO_2 without using the energy released for anything at all. Maybe, thought Ms. Glazer, that free energy could be used onsite to power wastewater cleanup technologies that normally wouldn't be considered because of their high energy costs. It also occurred to her that since lots of these wells are in the sunny, windy southwestern US, local photovoltaic panels or wind turbines might supply energy as well. This latter option is attractive when there are no convenient transmission lines to take the power offsite, even though solar or wind energy is abundant.

So the first idea is to channel the heat from burning the otherwise flared gas to heat-powered water treatment technologies like multi-stage flash distillation, multi effect distillation, and mechanical vapor recompression. Heat could also come from a solar thermal facility. Alternatively, if not enough gas were available for producing the heat, electricity from photovoltaics or wind turbines could pressurize the wastewater for membrane separation, or the related reverse osmosis. Reverse osmosis is the most energy-efficient of the treatments considered here, so it would be preferable if energy were limiting.

Ms. Glazer uses a series of equations to do a formal engineering analysis of the feasibility of her suggestions, and figures they could reduce overall water requirements for fracking by 11–26%, and reduce the energy use for freshwater trucking by 16%. If renewable energy were used there would need to be at least four 100 kW wind turbines (much smaller than the ones we are accustomed to seeing at wind farms) or between 1000 and 4000 250 Watt PV panels. She thinks she's on to something, and I'd agree. But even though the fuel is free, the equipment and maintenance surely are not, so this probably won't happen without a mandate, and unless there's a full-scale demonstration of the feasibility and cost, no mandate is likely either. Maybe DOE or EPA should fund a demonstration project.

Using Supercritical CO_2 Instead of Water for Fracking

The purpose of hydraulic fracturing is to use high pressure to open up pores in deep fuel-bearing shale deposits so that the oil or natural gas can escape through boreholes to the surface. To make this work, very high pressures (hence, much surface equipment) and a great deal of water are required. To keep the pores propped open when the pressure and water recede, something (usually sand) needs to be included. The inclusion of acid can increase pore efficiency, and because water is a good biological medium, antibacterial agents may be required to prevent fouling. Finally, most of the fracking fluid returns to the surface where it presents a treatment and disposal problem. But in theory, any liquid, or supercritical substance, would work, supercritical CO_2 (s CO_2), for example. According to a study underway at Los Alamos National Laboratory (Middleton *et al.* 2014) s CO_2 has a number of potential advantages over water, and some potential disadvantages as well.

The advantages are striking; it requires less pressure (so less equipment) at the well pad, it displaces gas from lower-porosity fractures and mobilizes it from organic inclusions, it mobilizes heaver hydrocarbons, it doesn't need the additives—maybe not even the proppant—now included in fracking fluid, and, perhaps best of all, it competitively displaces methane from the shale, preferentially absorbing on to it, staying in the shale rather than returning to the surface. That is to say, it is sequestered in the shale; just the thing we need to minimize releases of CO_2 into the atmosphere.

On the other hand, it costs more than water, and there is little information on its ultimate fate and on how to separate the volume of it that does return to the surface from the hydrocarbons.

The preliminary results of the study, all carried out using computer models based on known physical principles, are encouraging. Particularly encouraging is an effect seen in the s CO_2 and not in water: an abrupt cooling of about 200°C when the pressure is initially released, further shocking and potentially further enhancing crack propagation.

So, here we have a potential win-win situation; an economic reason to inject CO_2 under pressure into shale fields where it may be sequestered, and an increase in the effectiveness of oil and gas production by fracking with potentially less impact than the current hydraulic practice. Time will tell.

Heavy Oil Production using Fracking and Microwaves

Fracking isn't just for natural gas and conventional oil; it also increases the production of heavy oil in low permeability reservoirs, but if the oil is heavy enough and the cracks don't penetrate very far, the flow rates decay rapidly and not much oil is recovered. Heavy oil, not defined in this paper by Davletbaev *et al.* (2014), is usually a term used for oil just a little less viscous than bitumen, the "extra heavy oil" found in the highly contested Canadian tar sands—more or less like asphalt. One possibility is to heat up the rock surrounding the well to make the oil less viscous. A technique that works in conventional oil wells is to inject steam, but with heavy oil in low permeability reservoirs steam doesn't work very well. The approach discussed in this paper is to use microwave radiation (also known as radio-frequency electromagnetic radiation) of the sort used in your kitchen microwave oven. Of course it is impossible to bring the oil-shale to the oven—the oven has to go to it in the form of downhole electrodes. Another difference is the amount of energy required. A "powerful" home microwave oven consumes about 1,000 Watts (1 kW) provided by a standard kitchen electrical circuit. The heating of well bores simulated in this paper used 10–30 times that much electricity, but experimental studies have shown that after a day-and-a-half of heating, temperatures in the well can exceed 300°C (572°F) and can raise the temperature of the shale (and oil) to over the boiling point of water a few meters away.

This paper is about an operational strategy to maximize the flow of heavy oil using a combination of fracking and microwave heating, while minimizing the amount of electricity required. The idea is to frack first and let as much oil out as will come, then to heat for a while, stopping when the oil starts flowing, and reheating as many times as it takes to get as much of the oil as economically possible. The study compared the simulated oil production from a "cold" fracked well with that from a heated one for 550 days, assuming that the microwave device cost $100,000 and the oil sold for $100 per barrel. The authors concluded that the multi-stage heating could increase oil production by 87% in a well with low permeability "short" fractures, the type of well most suitable for this technique. Depending on the permeability of the fractures, the oil production rate, the price of oil, the amount of electricity used, and few other variables, payback times for the microwave heating ranged from 420 days to five-and-a-half years; somewhat less than the 550 days the simulations were run to four times that. The authors don't comment on what this says about the feasibility of using microwaves to increase heavy oil production, but from my calculations, it doesn't look like a very good option for most of the cases they tested.

Fracking: Fix it or Forget It? Global Gas and Oil Prices Falling.

Daniel E. Klein, an energy industry consultant, writes an interesting piece about fracking problems in Natural Gas & Electricity, an industry newsletter. His approach is to look at the prognostications of the Energy Information Administration Annual Energy Outlook (AEO)—pretty much the bible of energy projections—as they have changed from 2000 to projections of where we will stand in 2040. For example, there wasn't much shale gas until 2005 and in 2005 the AEO predicted that US natural gas imports would increase sharply in the near future. The 2014 projection, however shows the opposite: a steady increase in US exports, at least through 2024. Similarly, "peak oil" in the US has also been reversed by shale oil production, with the crude oil production in 2013 the highest in 25 years, and imports falling sharply, at least so far. OPEC has been debating, on the one hand, decreasing oil production, so as to increase global oil prices and therefore revenues or letting production stand so as to lower prices even further to put price pressure on American fracking operations. For now they have settled on the latter, but in the short-term oil prices will have little effect on American oil operations.

A few years ago the AEO projected that natural gas would only ever constitute 10–20% of fuel for power generation; now it seems to be displacing coal sufficiently to account for 25% now, and 30% by 2040. This has led not to "peak oil", but to "peak CO_2". It looks as if we are on a permanent CO_2 emission downslope in the US without ever having had to bite any bullets; we are leading the world in CO_2 reduction purely because the price of shale gas in the US has made it competitive with coal. When you hear the politicians from coal-producing states blaming the Obama administration for their troubles, it is therefore largely disingenuous.

The bottom line of this article is that whatever the environmental problems with fracking, they can be fixed, and the benefits are so monumental that they ought to be. I imagine that this is more-or-less correct, but there's probably nothing as disconcerting as having the bucolic farmland that once surrounded you turned into a heavily industrialized oil field—that's almost impossible to fix. However, unless one is constitutionally opposed to fracking, one ought to hope that it can replace coal globally with time...at least long enough for renewables to come of age.

Fracking in South Texas: Spatial Landscape Impacts

In a Master's thesis from the University of Texas at Austin, Jon Paul Pierre presents an interesting analysis of the effects of development (which includes a good deal of horizontal drilling and hydraulic fracturing) in the Eagle Ford Shale play in South Texas, where more than 5,000 wells have been drilled since 2008. What he sets out to do is assess the spatial fragmentation of the landscape from the construction of drilling pads, roads, pipelines, and other infrastructure. He used 2012 aerial photography with a 1-m resolution obtained from the National Agricultural Imagery Program (NAIP), and over laid on that the locations of well pads, pipelines, and other infrastructure, then used Geographical Information System (GIS) tools to characterize the types of areas being disturbed.

For the 628 wells in La Salle County (a portion of the overall study area) for which there was evidence of associated infrastructure on the aerial imagery, pipeline disturbance occupied 97 square kilometers, and drilling pads, 17 square kilometers. These activities heavily disturbed 3% of the county area, but 8.7% of the core areas, with a reduction in overall vegetated area from 91% to 89%, and of forest area from 76% to 68%. Probably of more concern than the absolute loss was the ecological spatial fragmentation caused mainly by the pipelines, potential soil loss from wind erosion of the disturbed areas, and interference with normal drainage patterns.

There aren't any particularly novel conclusions from this research, but it is a good example of how to analyze the physical effect of large-scale well development (or any kind of spread-out industrial development) on the landscape. Nevertheless the landscape is pretty thoroughly disrupted by oil and gas development, maybe even more so than this thesis implies.

Try Not to Live Too Close to a Fracked Well

If you happen to live within 1 km of a hydraulically fractured well in Pennsylvania, and you get your water supply from a well, you stand about twice as large a chance of having skin and upper respiratory problems than if you live 2 km or farther away; you have over 3 health symptoms, on average—people further away have only 1.6. Looked at another way, 13% of people living near fracking operations have upper respiratory problems, versus 3% living farther away; and 39% of the same group of people have upper respiratory problems versus 18% living further away. That is the disturbing result of an epidemiological study of almost 500 people in an area of natural gas drilling in the Marcellus Shale, just published in a journal of the National Institute of Environmental Health Sciences (Rabinowitz *et al.* 2014).

A major limitation of the study is that these conditions were all self-reported, and it is within the realm of possibility that people living near gas wells were more sensitive to the possibility of health problems, and more likely to report less severe ones. On the other hand, people living close to the wells might be receiving revenue from them and be more hesitant to report health problems…nobody knows yet. Clearly there needs to be a follow-up study in which health professionals characterize the symptoms.

What caused the reported symptoms? It could be contaminants leaking into the water or air from the fracking operations; or it could be related to the increased stress and anxiety of living near them. This study raises more questions than it answers, but that's the way it is with science, and it certainly adds to the body of knowledge suggesting that fracking is not without serious issues and in need of better regulation.

Biotic Impacts of Fracking

How does shale-bed energy development, including hydraulic fracturing, affect ecology? There have been a number of studies looking into this, and a new review paper by Sara Souther at the University of Wisconsin and seven colleagues at a diverse array of other institutions summarizes the current knowledge and where the gaps are in it. Their legitimate fear is that substantial damage will be done before much is known about the issues, and there is plenty of experience with other rapid industrial development to warrant concern. As an example, consider the damming of nearly all the rivers on both the east and west coasts of the US with little attention paid to the consequences for salmon.

The big issues they identify are subsurface and surface water contamination by fracking fluids, diminished stream flow because of water diversions for fracking, habitat loss and fragmentation, general disturbance to wildlife from the noise, light, and air pollution of fracking operations, and, of course, the atmospheric increase in greenhouse gases resulting from both leakage of natural gas in the process of collecting it, and CO_2 when the gas is burned by end users.

As the authors correctly point out, most of these are difficult to study, not much studied yet, and not evidently being studied to the degree necessary to properly evaluate the impacts. They'd like to see that change, since it doesn't appear that oil shale development is likely to slow down any time soon.

Methane Emissions in Colorado Exceed EPA Estimates; Fracking?

Colorado's north Front Range, north of Denver and east of Boulder and Fort Collins has become a frackers' paradise, with 24,000 active wells in 2012, 10,000 of them drilled since 2005. In the hot muggy summers, volatile organic compounds, including methane, ethane, propane, butane, pentane, and sometimes the carcinogen, benzene,(all commonly found in oil and natural gas (O&G), accumulate in the air, leading to elevated ozone levels, and contributing to global warming. Previous estimates of the total amounts released were based on a combination of bottom-up estimates of releases from various sources based on a variety of sampling methods, as well as air samples from tower sampling stations. Extrapolating these to the whole O&G area carries all of the uncertainty associated with each of these estimates. In order to get a top-down, fully integrated estimate, Pétron *et al.*, research scientists at NOAA, sampled the area from an airplane equipped with an instrument that continuously recorded methane, carbon dioxide, and carbon monoxide concentrations, and was also capable of taking discreet air samples for measuring other volatile organic compounds typically released from O&G operations. They found that the concentrations of most volatile organic compounds were twice as high, and that of benzene was seven times as high as previously estimated by the state of Colorado, and the hourly emissions rate was three times as high as estimated by the USEPA. The bottom line is that a lot more methane and other volatile organic carbons are being released from the O&G operations than was previously thought.

The researchers made a systematic effort to factor out the non-O&G methane emissions from the many agricultural sources in the region, using county-level agricultural census statistics from Colorado and the USEPA to determine the head count for several categories of cattle, as well as for sheep and poultry. This they multiplied by standard methane emission factors for each type, assuming a 20% uncertainty in both head count and emission factors—that their estimated head count was within 10% of the permitted capacity supported its accuracy. They also estimated emissions from manure management, but with less certainty owing to a lack of information on the details of the individual management systems. Additionally they estimated releases from the municipal landfills and both municipal and industrial wastewater treatment plants. All of this resulted in a "bottom-up" estimate of methane emissions from all non-O&G sources, which allowed them to attribute the rest to the O&G operations. The aircraft also took flask samples to measure non-methane hydrocarbon concentrations, and all of these were linearly correlated with the methane concentrations, suggesting the same source, presumably O&G operations.

It isn't clear why there is such a large discrepancy between bottom-up and top-down estimates of O&G-related volatile organic compound releases, but a reader of this paper would infer that there is not nearly enough monitoring of ground-based releases to give any sort of accurate overall results. There are as many as several thousand new wells being drilled and hydraulically fractured each year in their study area but, according to the authors, only a subset of them is subject to inspection; if they are drilled by the largest of the over 100 companies in the area that subset is inspected every 3 years; if they belong to a smaller company, only every 5 years. Clearly there is inadequate ground-based monitoring going on if a comprehensive view of O&G releases is desired.

Are any of these releases attributable to fracking? Virtually all of the new wells are fracked, but this study design had no way of determining at what stage of the production cycle the releases occurred.

Unexpectedly High Methane Concentrations over Pennsylvania Shale Gas Fields Too

Methane, the main constituent of natural gas (both that from gas wells and from farm operations) is a powerful greenhouse gas, around 30 times more potent than CO_2 over the hundred years after it is emitted. It is on the rise in the air above Pennsylvania, and the culprit might be shale gas development, which utilizes hydraulic fracturing. Caulton *et al.* (2014) used an airplane to sample the air above a 2,800-square-kilometer area of the Marcellus shale formation gas fields in Pennsylvania. It was rich in methane, with between 2 and 15 grams heading skyward over each square kilometer every second, the upper limit of which is quite a lot higher than the 5 grams estimated from what was previously known about wellhead methane emissions; the authors suspected that the transient nature of gas leakage might be the reason, making very difficult to come up with an average over time from ground-level measurements. Since they were in an airplane, however, they could circle around areas of high concentrations and pinpoint the source. It turns out that, sure enough, the sources were well pads and, in one case, a coal mine, but the interesting thing is that these wells were in an early state of development, hadn't reached their full depth, and hadn't yet been hydraulically fractured. The large releases at this stage of development were completely unexpected, but the culprit pads were visibly missing commonly used approaches that might have prevented the releases (shale shakers, mud pits), so it may have been carelessness on the part of the drillers causing the releases.

Similarly high measurements have been obtained from flights over well fields in Texas, Oklahoma, and Kansas, suggesting a widespread national problem. At this stage it doesn't look like the problem is fracking, so much as failure to employ standard techniques to capture the methane and save it or flare it, but clearly something needs to be done.

How Long will the Fayetteville Fracking Play Last?

How long will shale gas be available until it plays out? The Bureau of Economic Geology (BEG) at the University of Texas at Austin is making a concerted effort to find out for the four largest shale plays currently in development in the US. The first they reported on was the Barnett Shale in Texas. The topic of this post is their second study, conducted on the Fayetteville (Arkansas) Shale by John Browning and eleven colleagues at the BEG. The overall answer is a long time—but well short of a century—with production peaking soon and falling to between half and a third of the current levels by 2030 and continuing to decline thereafter; they ran their model through 2050 and estimate the technically recoverable gas resources if economics were not an issue (38 trillion cubic feet), and the amount likely to be recovered eventually given economic reality, about half that.

The article. Published in the Oil & Gas Journal (Browning *et al.* 2014) gives insight into the complex modeling required to come up with a credible answer. The authors looked at the gas production history of every (3,689) well drilled in the basin up to 2011, mapped their estimated 30–year productivity in a fine grid across the 2,737 square mile study area, extrapolated that productivity to nearby areas, and concluded that only about a third of the economically feasible wells have yet been drilled. As one would expect, all of their estimates are highly dependent on assumptions entered into the model, perhaps the most uncertain of which is the price of natural gas—if it is high, many more wells will be drilled than if it is low, but the model can accommodate just about any likely variation, so as more information appears over time, it will be a snap to rerun it.

I recommend this paper for its clear treatment of a difficult subject. For an example of a much more technically challenging description, try reading Patzek *et al.* (2013) by three of these same authors.

Conclusions

Depending on one's outlook, hydraulic fracturing is either the most stupendous windfall of energy experienced by the US in quite a long time, just in time to save us from peak oil and pushing peak coal and it's massive CO_2 releases to the past tense, or, an environmental disaster of the first water, destroying landscapes, delaying deployment of renewable energy technologies, continuing large, if diminished, releases of CO_2 into the foreseeable future, contaminating potable drinking water, and disrupting rural economies, lifestyles, and health.

To some degree this is a glass half full or half empty problem; it surely is both of these things to some degree. It will be some time before the definitive cost/benefit analysis is written and we know for sure, but though it may be delayed or derailed in some jurisdictions, it is unlikely to be stopped because it has the twin properties of improving the US economy and energy supply in ways little else could in the short term, and it has and will continue to decrease CO_2 emissions over those of the coal it has replaced. But as this chapter shows, there remains much research to be done.

References Cited

Bazant, Z.P., Salviato, M., Chau, V.T., Viswanathan, H., Zubelewicz, A., 2014. Why Fracking Works. Journal of Applied Mechanics 81.

Browning, J., Tinker, S.W., Ikonnikova, S., Gülen, G., Potter, E., Fu, Q., Smye, K., Horvath, S., Patzek, T., Male, F., 2014. Study develops Fayetteville Shale reserves, production forecast. OIL & GAS JOURNAL 112, 64-+.

Caulton, D.R., Shepson, P.B., Santoro, R.L., Sparks, J.P., Howarth, R.W., Ingraffea, A.R., Cambaliza, M.O., Sweeney, C., Karion, A., Davis, K.J., 2014. Toward a better understanding and quantification of methane emissions from shale gas development. Proceedings of the National Academy of Sciences 111, 6237-6242.

Davletbaev, A., Kovaleva, L.A., Babadagli, T., 2014. Heavy Oil Production by Electromagnetic Heating in Hydraulically Fractured Wells. Energy & Fuels. Just Accepted Manuscript DOI: 10.1021/ef5014264.

Glazer, Y.R., 2014. The potential for using energy from flared gas or renewable resources for on-site hydraulic fracturing wastewater treatment. Thesis, Master of Science in Engineering, Graduate School. University of Texas at Austin, 83 pages.

Klein, D.E., 2014. Fracking: Fix It or Forget It? Natural Gas & Electricity 31, 1-8.

Kurose, S., 2014. Requiring the Use of Tracers in Hydraulic Fracturing Fluid to Trace Alleged Contamination. Sustainable Development Law & Policy, Summer 2014, page 43.

Middleton, R., Viswanathan, H., Currier, R., Gupta, R., 2014. CO_2 as a fracturing fluid: Potential for commercial-scale shale gas production and CO_2 sequestration. Energy Procedia 63, 7780-7784.

Patzek, T.W., Male, F., Marder, M., 2013. Gas production in the Barnett Shale obeys a simple scaling theory. Proceedings of the National Academy of Sciences 110, 19731-19736.

Pétron, G., Karion, A., Sweeney, C., Miller, B.R., Montzka, S.A., Frost, G.J., Trainer, M., Tans, P., Andrews, A., Kofler, J., 2014. A new look at methane and nonmethane hydrocarbon emissions from oil and natural gas operations in the Colorado Denver-Julesburg Basin. Journal of Geophysical Research: Atmospheres.

Pierre, J.P., 2014. Impacts from above-ground activities in the Eagle Ford Shale play on landscapes and hydrologic flows, La Salle County, Texas. University of Texas at Austin, Master of Science Thesis, August 2014.

Rabinowitz, P.M., Slizovskiy, I.B., Lamers, V., Trufan, S.J., Holford, T.R., Dziura, J.D., Peduzzi, P.N., Kane, M.J., Reif, J.S., Weiss, T.R., 2014. Proximity to Natural Gas Wells and Reported Health Status: Results of a Household Survey in Washington County, Pennsylvania. Environ Health Perspect.

Souther, S., Tingley, M.W., Popescu, V.D., Ryan, M.E., Hayman, D.T., Graves, T.A., Hartl, B., Terrell, K., 2014. Biotic impacts of energy development from shale: research priorities and knowledge gaps. Frontiers in Ecology and the Environment 12, 330-338.

The Future of Tidal Energy

Cassandra Burgess

Tidal energy has the potential to be a major source of renewable energy. Currently, however, few functioning tidal energy converters exist, and development is slow to progress. Major economic and environmental obstacles stand in the way of large-scale development of tidal energy, but recent research shows that these challenges can be overcome. As it becomes apparent that fossil fuels need to be phased out quickly if we hope to prevent drastic changes in climate, removing these difficulties becomes even more pressing. While tidal energy remains severely underdeveloped compared to energy sources such as wind and solar, it has the potential to provide a great deal of energy. If tidal energy fulfills its promise, it can provide a renewable source of energy with low environmental impacts, and without the unpleasant appearance and land use difficulties that face wind and solar. In this sense, tidal energy is a much more ideal source than those currently used. It does not take away land that could be used for other purposes, and it does not have to disrupt the local ecosystem greatly if designed correctly. Tidal energy turbines are out of sight of most people, and when submerged, hardly noticeable at all. This discrete presence can help mitigate the "not in my backyard" effect that wind development has struggled with in some areas. Tidal energy is a huge, largely untapped source of energy that can be developed, but it faces two major obstacles.

The first major obstacle for tidal energy is efficiency. The devices must be efficient enough to produce large amounts of energy, while requiring little maintenance. Installing and maintaining devices in the ocean can be a huge undertaking, and in order for development to be possible, this undertaking must make economic sense. Recently, several groups have worked to develop accurate models of turbine outputs, as well as to determine the characteristics of turbine arrays that produce more energy. The design of a tidal energy converter array can have a large impact on its efficiency. According to Karimirad, Koushan, Weller, Hardwick, and Johanning, the design of moorings and foundations also has an impact on how much energy the array can produce. Many researchers are working to establish better models for the outputs of individual converters while minimizing their impact on turbulence and speeds around them. This research is important to understanding how turbines will interact with each other in arrays. Current models tend to underestimate speed and turbulence near turbines, preventing planners from understanding how an array of devices will actually function (Johnson). More recent models have become more complex in an effort to counteract this problem, modeling the turbines in more dimensions, and including more factors.

The other major obstacle for tidal energy is environmental damage. There are many concerns about how turbines will affect the organisms living around them, current speeds, tidal patterns, and many other aspects of the ecosystem. If installing turbines risks a massive environmental effect, their development will most likely not continue. It is difficult to argue that a new renewable energy source is viable if it will cause huge new environmental problems. Though many impacts are still unknown, researchers have recently begun tackling some of the questions about environmental impact.

Environmental damages come in two primary forms. The first is direct harm to living organisms near the tidal energy devices, particularly when they collide with turbines, resulting in serious injuries and death. This is particularly concerning when tidal energy devices are being installed in areas that have already been overfished. The threatened fish populations might experience further stresses if large numbers of their population are dying from turbine collisions. Other organisms such as diving birds and marine mammals face the same issue. As turbines are generally installed in areas with faster currents there is also the risk that animals using the currents for migration or hunting will face difficulties dealing with obstacles, and be unable to fight the current to avoid the turbine. The impact on animals largely depends on how quickly animals can react to the presence of a turbine. The second form of environmental damage is damage to the ecosystems. Increases in flow rates around the turbines may lead to increased erosion of the ocean bed. Bottom dwelling plants and animals depend on this habitat. Decreased current speeds in the wake of turbines may also cause fish to congregate near turbines for food. This can shift fish schools away from other natural habitat in the area, as well as draw larger predators nearer to the turbines, where they may collide with turbine blades.

Currently, the full effects of altering current patterns and velocities are unknown. However, several researchers have commented in their papers that there will likely be negative impacts. What these impacts will be is uncertain, as is their severity. Thus, it is difficult to set a limit on how much turbines should be allowed to alter the ecosystems they are in. It is also difficult to find ways to counteract the effects, or to predict in what regions the impacts will be most severe. These uncertainties are limiting the development of tidal energy converters throughout the world. Without being able to show that impacts will be minimal, or even able to definitively state what the impacts will be, developers are having difficulty installing new arrays. Research in this area is necessary and urgent. Before we can fully develop tidal energy we must understand its impacts.

Some researchers are optimizing for both efficiency and environmental impact. One new model, described in *A framework for optimizing the placement of current energy converters* (Roberts, Nelson, Jones, James), is designed to predict how an array of turbines should be arranged to both maximize energy output and minimize environmental impact. Models like this are important to the progression of turbine development, because they do not consider the issues separately, but rather create a comprehensive view of the turbines.

As the articles described in this chapter show, tidal energy is a viable future source of energy. Increasing efficiency can be done through small changes, such as using a different foundation, or a different configuration. While environmental impacts are not fully known, research is being done to determine what the impacts will be, and the results so far indicate that the damage will not be severe. The future of tidal energy is by no means certain, but recent developments look optimistic for its growth.

Impacts of Tidal Energy Converter Configurations

As tidal energy develops throughout the world several different designs for Tidal Energy Converters (TECs) have been developed. The main classifications are reciprocating and rotating devices, of which rotating are the most common. Within this category the TECs can be either floating and anchored to the bottom or fixed to the bottom by a rigid structure. Each of these designs has different impacts on the environment it is placed in. Sanchez *et al.* (2014) tested these impacts in Northwest Spain in Ria de Ortigueria by using a three-dimensional model to examine the impacts of two plants, one floating and the other bottom-fixed. The researchers found that both plants had little effect on water more than 4 kilometers away, but large impacts near the plants. The plants also exhibited very different patterns in flow near the plants. Thus the authors argue that because TECs clearly have an impact on the flow of water around them, further investigation will be necessary to find the impacts of this change in flow on ecosystems.

The research focused on two plants in this region. The estuary in the region experiences the strong current and water depth that are ideal for tidal energy generation, and follows a tidal cycle characteristic of much of Europe, making it a good indicator for possible environmental impacts of TECs in other areas of the world. Both plants were the same size and had the same mechanical features. One was floating and placed in the upper half of the water column while the other was bottom fixed and placed in the lower half of the water column. The Delft3D-FLOW model was then applied. The model spanned an area about 10km away from the plant, and was verified by testing it on actual data from 2010. The model was run without the TECs present, with a floating plant present, and with a bottom-fixed plant present. In each run the boundary conditions were forced to maintain a typical sea surface level and flow rate of the river.

The model predictions indicate that while both plants created small (1/100 meter/second) reductions in flow rate on areas far away, the floating plant created a large decrease in the flow rate at the surface that wasn't present in the floating plant simulation. Both designs caused an increase in velocity south of the plant in all depths of water. The model also predicted that the floating plant would produce approximately 1000 MWh more than the bottom-fixed plant, because of differences in flow rates at different water depths.

The results indicate the tidal energy plants can have large impacts on flow rate in areas near them, and these impacts vary significantly with the design of the plant. The impacts can also reach unexpected places, indicating that modeling should be done to determine the effects of plants before they are built. Depending on the ecosystem the plant will be built in the least impactful model and positioning can then be selected. In order to fully understand the environmental impacts of tidal energy plants, further study must be conducted on how variations in flow, like those the study describes, affect organisms living in the region.

Accuracy of Models for Wind and Tidal Turbines

Computational Fluid Dynamic models are used to investigate the influence of rotating wind and tidal energy generator turbines on the surrounding environments. Johnson *et al. (2014)*, compared current analytical and numerical models and experimental findings to a new computational fluid dynamic (CFD) model, and found that the CFD model agreed well with a simple conservation of momentum model, but did not closely match the experimental and numerical findings on reactions to the spinning turbines. This result was especially pronounced far from the turbine. The numerical and experimental findings predict much more turbulence downstream from a turbine, and larger changes in velocity.

This research is meant to serve as a benchmarking study to determine the accuracy of models and the best model to use in a variety of situations. However, this study is limited to a specific category of turbine. It assumes incompressible flow, implying that the fluids in question are moving at relatively slow speeds. This assumption is most likely to be valid for any wind or marine situation, as the upper limit of Mach 0.3 indicates an upper speed limit of 102 meters per second. All of the systems considered are moving at speeds far below this. The model also assumes an open channel with no other obstacles disrupting flow. Finally, the discs that compose the turbines were modeled as infinitely thin and as having a constant load applied. This does not match the conditions of the real world.

The major differences between the models studied and the experimental data occurred in the definition of the changing momentum due to the turbine. In the models this was defined to be constant, and in only one direction. Because of this the model tended to predict a lower velocity and turbulence than the experimental data, particularly in the region near the turbine. The models also predicted that turbulence peaked further downstream than the experimental data. Based on these results, future analysis of turbine environmental impacts must take into account that the most common models underestimate the resulting turbulence. The difference in predicted turbulence and location of turbulence could have significant impacts on the optimal placement of turbines.

Marine Energy Device Mooring and Foundation Designs

The deployment of marine renewable energy (MRE) devices requires new approaches to offshore mooring and foundations. Mooring and foundations are essential components of MRE device structures, enabling a device to remain in one spot while responding to the movement of tides or waves. Foundations and moorings often account for up to 10% of the total cost of building and maintaining an MRE device; designing a foundation of mooring that can be built at minimal cost requires analyzing the site characteristics, the direction and magnitude of forces on the device, and responses of different technologies to these forces, and the ease of installing and decommissioning the device. These steps present a novel challenge for MRE devices, because such devices have different requirements than traditional offshore technologies. One of the main challenges in applying traditional technologies to this problem is that MRE devices often need to be able to move. This is because the response of the device itself to tides and waves, which is necessary for energy production, is generally linked to the response of the foundation or mooring system. Thus these systems are required to endure loads both horizontally and vertically and to move with the device in some cases. This differs from more common offshore technologies, such as drilling platforms, where movement is not desired. Because of this novel challenge, Karimirad, Koushan, Weller, Hardwick, and Johanning suggest in their 2014 paper that new guidelines need to be developed to aid in the design of MRE device structures.

In addition to the concerns with different loading needs, MRE devices also present challenges in design reliability. These devices are likely to be built in arrays in order to make them economic and to facilitate easy installment and maintenance. However in arrays there is a new concern about possible tangling of mooring lines. The authors analyzed several mooring systems, such as taut lines, catenary lines, and surface buoys. In designing these systems it is often desirable to include multiple lines in case one should fail. However, having multiple lines increases the likelihood that the devices will tangle with each other, which could lead to failures in lines, as well as collisions between devices. In addition catenary lines, which are not taught, are often a better choice because their design lowers the peak loading, allowing the mooring to function under more conditions without failing. This is important when considering offshore devices, because the moorings must not fail in the event of large waves that cause greater than normal stresses on the system. Because the lines are not taught, however, there is a possibility that they will become entangled with neighboring lines. These concerns led the authors to consider surface buoys, and different foundation arrangements, in order to handle the increases stresses without raising the risk of line entanglement. Their paper provides a summary of the advantages and disadvantages of each technology examined, and ultimately concludes that new construction guidelines accounting for the increased complexity of MRE device mooring and foundation design are necessary.

Low Density Tidal Energy Arrays Minimize Impact

The configuration of a tidal energy array partially determines the level of environmental impact. In determining the optimal configuration for a particular area, it is important to consider not only power output, but also environmental impacts. Fallon *et al.* discuss the impacts of a tidal energy array located in Ireland in their 2014 paper. They use a two dimensional model and average speeds over the depth of the channel. This simplifies the modeling process, but it also overestimates some of the impacts. They then analyze grid spacing for turbines, with turbines spaced 0.5, 2, and 5 times their diameters apart. The model indicated that the 5-diameter spacing had the least environmental impact. It decreased velocities by 19.9% less than the 0.5 diameter spacing outside the grid, and increased flood velocities by 27.3% less. The 5-diameter spacing also changed the tidal range of heights by only 1% while the 0.5 diameter changed them by 6.4%. Because the 5-diameter spacing has significantly less impact on the hydrodynamic environment around the turbines, the authors conclude that it is desirable to use low density arrays when possible.

The authors discuss the possible environmental impacts of changes in the hydrodynamics of the channel. Increased velocities around the array will likely lead to increased erosion of sediment in the bed and shoreline. This leads to greater amounts of sediment in the water, which increases maintenance costs for the array. The increased erosion will also likely impact those species living in the bed of the channel negatively. Changes in velocities impact many marine mammals and fish, and it is possible that these species will migrate to areas with more moderate tidal flows in response. While it is unknown what the exact effects of any changes in tidal currents and height ranges will be, they will certainly have some effect. It is important to minimize these effects wherever possible, without sacrificing the efficiency of the tidal array. This can be accomplished by building a low-density array with high power output. The low density arrangement dramatically reduces the environmental impacts, allowing the turbines to take more energy from the tidal flows without significant damage to the surrounding environment. The authors conclude that a spacing of 5 diameters is sufficient in this particular instance, but urge site-specific study for future arrays.

Optimizing Tidal Energy Converters

In order to make tidal energy converters economic enough to compete in the energy market it is essential to build them as efficiently as possible, but also important to design them to avoid environmental impacts on the habitats in which they are installed. These impacts can be more difficult to predict when planning an array of tidal energy converters than a single turbine. Roberts, Nelson, Jones, and James worked to solve these problems by creating a modeling framework that optimizes the placement of tidal energy converters in Cobscook Bay, Maine. The model uses restrictions on water height and velocity based on the region so that it can be applied to other regional sites as well. It also allows researchers to input environmental restrictions on the decrease in velocity due to the turbines, and on changes in the bed shear stress at the site. These constraints represent points at which the turbines might change fish behavior by causing fish to congregate in the turbine wakes, and at which erosion of the ocean floor becomes serious. Using these restraints the researchers found that the non-environmentally constrained system had an output 19% higher than the originally planned placement, and the environmentally constrained system had an output of 16% higher.

For the purposes of this modeling process the environmental constraints were set arbitrarily. In future models, research would be necessary prior to the planning of the tidal energy converters to determine what levels of change the ecosystems could reasonably withstand. Once this is determined, the model can optimize the placement of tidal energy converters while minimizing the environmental impact. This model differs from previous models because it is on a much finer scale. While previous models have been able to accurately predict the impacts of tidal energy converters on a broad scale, this model looks at the fluid dynamics near the turbines themselves. This improvement allows for analysis of the environmental impacts near the turbines, as well as for better information on the turbulence and velocity changes created, both of which affect the power output of nearby turbines. Because this model was able to optimize both energy output and environmental impact, two areas most concerning when constructing a tidal energy array, the researchers recommend that it be used in the planning for future array sites.

Fish Behavior near Tidal Energy Turbines

Any man-made structure in a marine environment has the potential to impact the organisms living there. Previous research has shown that fish actively avoid trawlers and boats, and that abandoned oil platforms often become a place for fish to congregate. It has yet to be determined how fish will react to the presence of tidal energy generators. This may be an important design consideration. Viehman and Zydlewski (2014) conducted research on fish behavior at a turbine in Cobscook Bay, Maine. This bay is known for high biodiversity, and provided the chance to study a range of fish species. The study found that over 50 percent of the fish they monitored in the area interacted with the turbine in some way, and that 34.8 percent were observed to enter or exit the turbine during the 22 hours study. They also found large fish (greater than 10 cm in length) were more likely to avoid the turbine at night than small fish. At night small fish had only a 0.002 probability of avoiding the turbine, while large fish had a probability of 0.109 of doing so.

This study was conducted using two DIDSON acoustic cameras. One tracked fish upstream of the plant, and the other tracked fish downstream. Unlike a conventional camera these allowed for tracking at night without using light, however, the range and resolution of these cameras is limited. Their greatest resolution was parallel to the current, so they were unlikely to detect fish moving at an angle to the current.. They also only tracked an area about 10 m away from the turbine. The study lasted for 22 hours, including 11 hours in daylight and 11 hours at night. This period constitutes two tidal cycles, allowing for observations of both high and low tide during day and night.

The researchers found that turbine rotation greatly affected fish behavior, reducing the probability of fish entering by 35 percent, increasing the probability of fish avoiding the turbine by 120 percent, and increasing the probability of fish passing by the turbine to increase 97 percent. In this study avoidance was characterized by the fish entering the camera view on a course to enter the turbine, but then changing course to avoid doing so. Passing by was characterized as entering the camera view either above or below the turbine and continuing past it. It impossible that some of the fish that passed by the turbine actually exhibited avoidance, but did so out of range of the cameras. These are very large differences in fish behavior based on the motion of the turbine.

Based on these results, it is clear that day or night conditions, turbine motion, and fish size all have an impact on how fish will behave near a turbine. These factors should thus be taken into account in deciding where to place future turbines, and at what times they should operate. Previous studies have shown that fish of different sizes move to different parts of the water column throughout the day, so the height and time of operation of a turbine could have a large effect on what species of fish interact with it. This study was small, spanning only a 22 hour period, and monitoring a small area around the turbine. Fish behavior outside of the 10 meter radius could not be tracked, and so any fish avoiding the turbine from a greater distance were simply noted as passing. Because of these limitations of the study, and the importance of understanding how fish interact with turbines, the researchers believe that a larger study should be conducted to better define which parameters most influence fish behaviors.

Diving Seabird Interaction with Tidal Energy Turbines

Tidal energy turbines are deployed in environments with many different species. The interactions between organisms and turbines can greatly alter the ecosystem, so it is important to understand what these interactions will be before installing the turbines. This is especially important when turbines will be installed in areas with already threatened or endangered species. In a 2014 paper Williamson, *et al.* describe how the FLOWBEC Seabed platform can be used to understand interactions between turbines and diving seabirds. This device measures current, chlorophyll, and turbidity in the region of a turbine along with acoustic tracking of mammal, fish, and bird movements to analyze the relationships. Because these relationships vary greatly between species and region the authors do not classify a specific relationship, but rather describe a framework for future researchers to use in analysis.

The researchers completed five trials of 2-week deployments in the UK. Each trial tested behaviors both with and without turbines. After taking the data, they classified the animals observed in the sonar system using size, shape, and number of animals. They also used multi-frequency analysis to verify and improve on classifications. This analysis notes that the frequency response of many fish species is known, or can be tested. The frequency response is essentially the fixed function by which the amplitude and phase of the returning sound wave are related to the amplitude and phase of the outgoing sound wave from the platform. Because the amplitude and phase of the wave output by the FLOWBEC platform are known to the researchers, measuring the amplitude and phase of the wave that returns allows the researchers to calculate the frequency response function. The sonar system can then be trained to check for this frequency response, and to then track the animals that produce it. This technique could be helpful in future analysis using this model, as it allows researchers to locate and focus on a specific species of interest, rather than investigating the entire range of species. It also provides more accurate differentiation between species, without requiring visual methods that can disturb the environment by introducing extra light. Using the data they gathered the researchers then performed statistical analyses to determine how behavior of the organisms present was correlated with the other factors they measured. This data then dictates where the turbines should be placed, and when they should be running in order to minimize the damages to marine organisms. The researchers state that this model is accurate enough to prevent population declines after turbine deployment. This will aid in the development of tidal energy technology, because it mitigates some of the environmental risks involved in the technology.

Risk of Collisions Between Turbines and Fish

The presence of spinning turbines in the marine environment introduces a new risk for the many species that interact with them. Marine mammals, diving birds, fish, and other organisms can be killed or injured when they collide with spinning turbine blades. The probability of this occurring has not been extensively studied, so it has been difficult to predict the impact that tidal energy turbines will have on marine populations. In their 2015 paper Hammer *et al.* develop a probabilistic model for the collision of fish with turbines, providing a quantitative way to make predictions of the impact. While the parameters of the model vary with the specific site and species of fish in question, the authors hold that the model itself is general enough to apply across many different areas. The model is a simple fault tree, which calculates the probability that a fish will be fatally injured by a turbine using the probabilities of a series of events. These events include array passage, co-occurrence, avoidance failure, size of the hazard zone, chances of fatal turbine injury, blade damage. Each of these factors depends on the characteristics of current flow in and around the turbine array as well as behaviors of the particular fish species being examined. Using this model for their analysis the authors conclude that turbines will pose a very low risk for small fish, but bigger fish have a high probability of collision, particularly at night when visibility is low.

The fault tree that makes up the model is simply a diagram of what events must occur in order for a fish to have a fatal collision with a turbine. Each of the probabilities necessary for the fault tree was examined by the researchers. They found that array passage, the probability that fish is in the same area as the turbines depended on the size of the turbine array as well as the range of the fish species. Co-occurrence, the probability that a fish is in the region while a turbine is in operation similarly depends on when fish happen to be in the area, and when the turbines are on. Avoidance failure is the chance that a fish in the region will fail to avoid the turbine. This can occur if the fish does not react to the presence of the turbine soon enough, or if the current is too strong and the fish exhausts itself trying to swim away from the turbine.

The hazard zone is the probability that a fish being swept into the turbine will be swept into the section of fast moving blades, rather than a slower, less harmful section. Turbine injury consists of the chances of collision with a bad and the chances of experiencing hydraulic stress due to pressure changes in the water. Finally, blade damage is the probability that a collision causes severe damage, given that one took place. Clearly each of these parameters depends greatly on the circumstances of the site and the behavioral habits of the fish involved. However, this model provides a framework for future research, and points researchers towards which questions they must answer in order to get an accurate understanding of how a turbine array will influence a marine population. Based on how each parameter varies with size, the researchers found the larger fish are more likely to be fatally hit by turbines. They are less able to slip through the blades unharmed, and more likely to be present in the strong currents that cause the turbines to move. Many smaller fish avoid these strong currents altogether and thus have a very low co-occurrence probability. From this the authors suggest that when future turbines are planned the exact parameters should be determined for any threatened or endangered species of large fish, so as to avoid decimating the population with the introduction of turbines. This study may not apply to other categories of animal, such as marine mammals, which have very different swimming habits than fish. However, the general framework is likely to apply, although the specific parameters and how they vary with changes in size, shape, etc., will be different.

Conclusions

Tidal energy has the potential to provide a huge amount of energy, reducing fossil fuel dependence and helping to counteract some of the forces of climate change. In order to achieve this tidal energy must be developed quickly. This development will be successful only if turbines can be made more efficient, resilient in the marine environment, and environmental risks can be overcome. Thus it is essential that research continue into the best designs for tidal energy converters. The problem of how to best harness the power of the tides is by no means solved. We do not know that impact turbines will have on many animals. We do not know what the most efficient design and arrangement of turbines is to optimize energy efficiency. We are making progress towards discovering these things, however. The papers summarized in this chapter represent some of the most recent research done on tidal energy. As long as this research continues to take place, tidal energy has a chance to make a huge contribution to the world's energy supply.

References Cited

Fallon, D., Hartnett, M., Olbert, A., Nash, S., 2014. The effects of array configuration on the hydro-environmental impacts of tidal turbines. Renewable Energy 64, 10-25.

Hammar, L., Eggertsen, L., Andersson, S., Ehnberg, J., Arvidsson, R., Gullström, M., Molander, S., 2015. A Probabilistic Model for Hydrokinetic Turbine Collision Risks: Exploring Impacts on Fish. PloS one 10, e0117756.

Johnson, B., Francis, J., Howe, J., Whitty, J., 2014. Computational Actuator Disc Models for Wind and Tidal Applications. Journal of Renewable Energy 2014.

Karimirad, M., Koushan, K., Weller, S., Hardwick, J., Johanning, L., Applicability of offshore mooring and foundation technologies for marine renewable energy (MRE) device arrays. The Norwegian Marine Technology Research Institute, DTOcean.

Roberts, J., Nelson, K., Jones, C., James, S.C., 2014. A Framework for Optimizing the Placement of Current Energy Converters.2nd Marine Energy Technology Symposium, April 15-18, 2014.

Viehman, Haley A., and Gayle Barbin Zydlewski. Fish Interactions with a Commercial-Scale Tidal Energy Device in the Natural Environment. Journal of the Coastal and Estuarine Research Federation (2014): 30 Jan. 2014. Web.

Williamson, B., Scott, B., Waggitt, J., Blondel, P., Armstrong, E., Hall, C., Bell, P., 2014. Using the FLOWBEC seabed frame to understand underwater interactions between diving seabirds, prey, hydrodynamics and tidal and wave energy structures, Proceedings of the 2nd EIMR conference, 28 April–02 May 2014, Stornoway, Isle of Lewis, Outer Hebrides, Scotland.

Analysis of Nuclear Power as an Electricity Generation Option

Cameron Bernhardt

Nuclear power has served the U.S. and California as a primary source of electricity for decades. Although considered a conventional fuel source, nuclear power offers significant environmental and economic advantages over many other generation types, with its near-zero greenhouse gas emissions from operations, high reliability as a baseload power source, and relatively low operating costs. Conversely, nuclear power generation often requires large amounts of water for cooling and large areas of land for waste disposal. Furthermore, the long-term air pollution and greenhouse gas emission advantages that nuclear generation offers are potentially offset by nuclear reactor meltdowns such as the Fukushima Daiichi disaster. A review of current literature on nuclear power generation emphasizes the role that trade-offs possess when evaluating the merit of nuclear power in case studies and the unique, diverse range of issues surrounding nuclear development.

Nuclear electricity production emits low greenhouse gas emissions and has often been supported for its capability to replace carbon-intensive sources. However, nuclear power has the potential to create other environmental issues, such as increased demands on water for cooling and land for waste disposal. Byers *et al.* (2014) tested a variety of decarbonization pathways to evaluate water usage in the U.K.'s electricity sector up to 2050. Out of the six pathways that were tested, those with the highest proportion of nuclear generation resulted in the highest projected tidal and coastal water abstraction. This relationship suggests a potential trade-off between reducing greenhouse gas emissions and reducing water supply security. Poinssot *et al.* (2014) reached similar conclusions regarding nuclear power's high demands on water withdrawal and added that nuclear waste management was an integral factor in achieving sustainability throughout the entire production cycle.

A wide range of policy responses to the Fukushima disaster are evident around the world. Some countries, including France, continued their development of nuclear generation, but countries such as Germany, Switzerland, Italy, and Japan indicated plans to phase out nuclear power. Bird *et al.* (2014) found that public support for nuclear energy in Australia was adversely affected in the aftermath of Fukushima. Furthermore, Akashi *et al.* (2014) demonstrated that the cost of reducing carbon emissions is likely to increase if nuclear power is phased out, but that material efficiency improvements could produce similar emission reductions even in the absence of nuclear generation. Nuclear power's role as a baseload electricity source make it difficult to be replaced by renewable energy, since renewables are typically intermittent and cannot be relied upon for consistent electricity generation. Vine and Juliani (2014) recognized this issue and advocated a diverse generation portfolio to balance the risks and costs of relying too heavily on one source. Real-time market price analysis by Woo *et al.* (2014) showed that a $6–9 per megawatt-hour (MWh) increase in regional wholesale electricity prices occurred from the closure of the San Onofre Nuclear Generating Station.

As mentioned above, the long-term treatment of nuclear waste is often one important issue to address in nuclear generation management. There are presently two fuel cycles and two respective waste disposal options that carry different management costs. The closed cycle strategy, unlike the open cycle "once-through" strategy, recycles uranium oxide elements for future use despite producing similar final waste products. Soria *et al.* (2015) examined the costs of two taxes on the two waste disposal processes in Spain and found that the environmentally superior closed cycle method was only 6.6 percent more costly, even in a "worst-case scenario." Poinssot *et al.* reached a similar conclusion and demonstrated the potential uranium conservation that could be achieved with a "twice-through" closed cycle. The differences between open- and closed-cycle fuel strategies suggest additional trade-offs to be considered between the economic feasibility of nuclear power and its environmental impacts.

This literature review considers a diverse range of topics surrounding nuclear generation and illustrates the need for further studies to assess nuclear energy's potential as a low-emission electricity source moving forward.

Nuclear Power Generation: High Demands for Cooling Water Use

Nuclear power is often praised for its potential to replace carbon-intensive energy sources and reduce greenhouse gas emissions from electricity. Although nuclear power may offer a promising future in this regard, it is likely to place stresses on the environment in other ways, namely through increased demands on water for cooling and space for waste disposal. Byers *et al.* (2014) tested six decarbonization pathways to estimate current water use in the UK electricity sector and project water use to 2050 in the UK. The study observed the water use associated with cooling for all varieties of thermoelectric power plants, but nuclear power accounts for over 20 percent of the UK's current electricity mix and is likely to share a large stake in the future of the UK's power mix. Byers *et al.* concluded that the pathways with the highest proportion of nuclear generation resulted in forecasted tidal and coastal water abstraction that exceeded current levels by up to six times. This finding suggests that nuclear power may often not be a viable future energy source, especially in areas where water resources are relatively scarce. It seems that the UK should extend its investigations into the merits of nuclear power, and similar studies may be warranted to assess the impacts of nuclear generation in other cases.

The study also emphasizes two related points that could be extrapolated to other scenarios. The first is that, despite the UK's relatively high access to seawater, evidence from the study indicates some scarcity of viable sites for large nuclear plants "if negative environmental impacts are to be avoided." This reality may be due to the UK's high reliance on freshwater as a thermoelectric cooling source. However, it was also found that significant reductions in freshwater demand are feasible through greater "hybrid cooling," which, although more costly and carbon-intensive, would increase the security of water supply. The study makes it clear that high levels of nuclear power generation create environmental risks related to water consumption that will require an analysis of trade-offs between costs, emissions, and the environment.

Environmental Impacts of French Nuclear Twice-Through Fuel Cycle

Currently, nuclear power generation provides France with approximately 75 percent of its electricity. Although this share is expected to be reduced to roughly 50 percent by 2025, it is likely that France will expand its nuclear capacity over the next decade in order to meet rapidly increasing energy demands. Nuclear energy provides France with a low-emission, stable, and relatively cheap source of electricity, as well as a significant percentage of the €3 billion that it gains from net electricity exports. Poinssot *et al.* (2014) utilized the Nuclear Energy Life Cycle Assessment Simulation tool, "NELCAS," to create a global footprint evaluation of the French nuclear energy system.

In the first part of their analysis, Poinssot *et al.* used the NELCAS tool to compare the environmental competitiveness of nuclear power to generation from coal, oil/gas, solar photovoltaics, hydropower, and wind. Results indicated that nuclear power in the current French energy portfolio has a relatively low environmental impact compared to other energy sources. Nuclear electricity production was demonstrated to produce less atmospheric pollution and greenhouse gases than these other sources. Conversely, nuclear ranked unfavorably in terms of water use, especially in water withdrawal. In consideration of these results, the authors added that nuclear waste management remained "the key critical impact of the overall nuclear cycle."

The second portion of the Poinssot *et al.* investigation compared the current French twice-through nuclear fuel cycle (TTC) to equivalent electricity production under an open, once-through fuel cycle (OTC) scenario. Results suggested that an OTC would have a relatively negative environmental footprint, primarily with regard to the non-radioactive generation aspects that were simulated. In addition, France's current TTC scenario allows greater than 17 percent savings in natural uranium resources. The analysis seems to support past and future development of TTC nuclear power generation in France as long as waste can be successfully managed, water issues can be avoided, and radioactive releases continue to contribute a relatively insignificant percentage of natural radioactivity.

The Aftermath of Fukushima: Public Opinion of Nuclear Power in Australia

Deciding the future of nuclear power generation significant for many countries around the world. Like all electricity generation technologies, nuclear power possesses notable advantages and disadvantages relative to other generation methods. Some of the most commonly recognized advantages of nuclear power are its low operating costs, security of supply, and the low air pollution and greenhouse gas emissions that it produces. Conversely, issues of water use and waste disposal are often deterrents to the development of nuclear generation. In addition, the risk of nuclear accidents is a persistent threat to nuclear support, especially after incidents such as the Fukushima Daiichi disaster in 2011. In an effort to characterize the Australian public's views toward nuclear power in relation to climate change and other alternative energy sources, Bird *et al.* (2014) analyzed random sample surveys to draw conclusions about these attitudes. These surveys were administered in March 2010 and February 2012, 12 months prior to Fukushima and 11 months following, respectively.

The results of the surveys showed that fewer Australians in 2012 viewed nuclear power as a "satisfactory option" for electricity production than in 2010. Furthermore, fewer were willing to accept the development of nuclear power stations even if such development would positively influence climate change. The percentage of respondents that believed the risks of nuclear generation outweighed the benefits increased from roughly 34 percent to 42 percent in the two-year span. These results suggest that Australian public support for nuclear power declined from March 2010 to February 2012, and it is likely that the Fukushima Daiichi disaster was a significant factor in this decline. However, the results from these surveys cannot explain what actually influenced perceptions of nuclear power; considering the roles that media and other forms of communication have on public attitudes may help further explain the decline in support.

High Costs for Achieving Emission Reductions Targets without Nuclear and CCS

Reducing greenhouse gas emissions from electricity generation is one of the primary means by which humans can mitigate global climate change. Employing nuclear power generation and carbon capture and storage (CCS) are two methods for decarbonizing electricity generation processes, but the merit of these technologies is often debated. While these technologies are often effective in lowering the amount of greenhouse gas emissions from electricity generation, they pose other environmental and economic threats that frequently limit their popularity and use. These realities have created some uncertainty regarding the future deployment of CCS and nuclear power generation. Akashi *et al.* (2014) use a multi-scenario analysis to investigate the feasibility of the international emissions reduction target (holding the increase in global average temperature below 2°C) in a future without nuclear or CCS technology. The authors considered four different scenarios: baseline, standard 50 percent reduction, 50 percent reduction with no CCS or new nuclear power plants being built, and a variant of the third scenario but with improved material efficiency.

Results warranted mixed conclusions regarding the importance of nuclear power generation and CCS for economically reducing greenhouse gas emissions. Akashi *et al.* found that while the emission reduction target is still technologically feasible, the costs for achieving the target increase greatly if nuclear power and CCS are limited. Conversely, results also indicated that material efficiency improvement measures such as recycling could curtail the cost of significant carbon emission reductions even in the absence of nuclear power generation and CCS. Specifically, a trade-off was identified between reducing greenhouse gas emissions in the energy sector and reducing emissions in the building and transportation sectors. With the future of two viable reduction methods undecided, there may be a need for "a major transformation of the electric and road infrastructure." Regardless, it is likely that reduction costs would increase significantly due to necessary new technology investments in the building and transportation sectors.

The Importance of Nuclear Power as a Zero-Emission Generation Technology

Although relying on nuclear power for electricity poses some environmental issues, nuclear generation does offer several notable advantages over other generation options. Vine and Juliani (2014) suggest that nuclear power's potential to produce significant amounts of electricity with nearly zero greenhouse gas emissions and to provide consistent base load power should not be overlooked. However, these benefits may not be sufficient to outweigh reinvigorated concerns about nuclear safety – the authors recognized that many nuclear reactors may be retired in the near future in both the United States and abroad. Four power companies in the United States alone have announced the retirement of five large reactors since late 2012, perhaps due to negative stakeholder perceptions of the risks that nuclear generation poses. The authors show that nuclear power's majority share in the U.S.'s zero-emission fuel sources will be difficult to replace should nuclear generation be phased out, especially with regard to its reliability as a base load power source.

Nuclear power contributed roughly 19 percent of total U.S. power generation in 2012 and accounted for more than four times the amount provided by wind and solar combined. Similar to other zero-emission generation types, a relatively large portion of source emissions come from building infrastructure. In this sense, any loss of nuclear generating capacity is likely to be uneconomic and result in increased greenhouse gas emissions. This is primarily due to the relatively high carbon emissions from establishing nuclear power infrastructure compared to typical operating processes. Assuming that a region retiring a nuclear power plant wants to maintain present levels of power generation, the region can either increase the share that carbon-intensive fuel sources have in its power generation or increase the share of renewables, both of which would require infrastructure expansion and also likely increase greenhouse gas emissions. Although it may seem like renewable development will always reduce carbon emissions, this is not necessarily true in the short term. Developing new infrastructure to operate renewable generation is carbon intensive and may not be offset by the carbon saved.

Electricity-Market Price Impacts from the San Onofre Nuclear Plant Shutdown

Since the Fukushima Daichii nuclear power plant disaster in 2011, the future of nuclear electricity generation has been brought into question. A wide range of policy responses to the Fukushima incident have been employed in many countries around the world, varying from nuclear expansion to immediate shutdowns of nuclear plants and the suspension of new plant approvals. In California, the San Onofre Nuclear Generating Station was shut down in January 2012 due to significant wear on over 3000 different tubes in the plant. This policy decision by the California Energy Commission (CEC) naturally had a huge impact on the state of the electricity market in California; the 2160-MW San Onofre plant provided a large share of the electricity to its surrounding region. In light of this CEC decision, Woo *et al.* (2014) analyzed the price impact of San Onofre's shutdown. Woo *et al.* used intra-hour prices to compute average real-time market prices from roughly 24,000 observations between California's three independent operating regions. The regression results led the authors to conclude that a $6-9/MWh increase in regional wholesale electricity prices occurred from the San Onofre shutdown. The authors also concluded that this price increase could be offset by reducing system load and expanding solar and wind generation, which would also likely decrease carbon emissions from electricity generation in California and mitigate climate change.

Woo *et al.* believe that their work provides the first analysis of the real-time electricity-market price increase resulting from the shutdown of the nuclear plant in San Onofre. Examining the policy implications of the San Onofre shutdown has global relevance due to the size of California's electricity market. The establishment of their regression model should further enhance the potential for analyzing market-price impacts from power plant closures around the world.

Nuclear Fuel Cycle Economics and Tax Impacts in Spain

The long-term treatment of nuclear waste is often considered one of the most significant issues to address when developing nuclear generation capacity. There are currently two fuel cycles and respective waste disposal options available to industrial scale nuclear generation: open cycle and closed cycle. The open cycle strategy functions as a "one time use" cycle where nuclear fuel elements are considered as high level waste and ought to be disposed of in a deep geological repository. The closed cycle strategy recycles uranium oxide elements for additional use, but the final waste products are still comparable to open cycle processes in the long run. Soria *et al.* (2015) used a comparative trend analysis model to quantify the costs of two taxes on used nuclear fuel in Spain for each waste disposal strategy. One tax applied to the production of used fuel when it was extracted from the reactor while the other applied to fuel after storage. While it was clear that these taxes would have significant impacts on the costs of back-end nuclear fuel management, Soria *et al.* sought to analyze the difference in management costs between open cycle and closed cycle disposal processes.

Results indicated that, under a scenario in which no credit was earned for reprocessing materials in the closed cycle, the closed cycle was 12 percent more expensive than the open cycle. While management costs for the closed cycle process were demonstrated to be higher, the cost of the taxes on the closed cycle was found to be 2.6 percent higher, thus reducing the difference in total costs between technologies. The tax burden, however, was found to be greater than the management cost burden, reducing the difference in total costs to just 6.6 percent, with lower costs for the open cycle process. Results from the study conclusively demonstrated that the open cycle disposal process was cheaper in terms of management costs and tax compliance, but it is important to recognize that these costs were estimated under a model that assumed a "worst-case scenario" for the closed cycle process, "where Spain does not receive any credits for the reprocessed material." Furthermore, the authors suggest that such a small estimation in cost differences may not be relevant due to variability in their model. They also highlight other distinguishing advantages between the disposal methods.

Economic Analysis of Fossil Fuel Subsidy Reform and Nuclear Phase-Out

Having established the many significant trade-offs to evaluate when developing nuclear generation plans, it also seems important to consider the impacts that a phase-out of nuclear power could carry. International leaders (the "Group of Twenty") adopted a declaration in 2009 to eradicate fossil fuel subsidies that encourage wasteful energy consumption. In developing countries, these subsidies encourage underpriced electricity, excessive burning of coal and natural gas, and petroleum consumption. It is likely that phase-out of these subsidies would improve social welfare and provide environmental benefits such as reduced greenhouse gas emissions. Furthermore, such policy reforms are likely to reduce dependency on fossil fuels and promote investment in cleaner generation options. Similarly, nuclear power generation is being phased out in several G20 countries after the Fukushima accident in 2011. Magné *et al.* (2014) used the OECD's general equilibrium model ENVLinkages to forecast mid-term economic implications of a simultaneous fossil fuel subsidy reform and nuclear power generation phase-out. The analysis was extended to estimate impacts on carbon emissions, assess interactions between the policy reforms, and measure the associated costs of each policy. Magné *et al.* concluded that an effective fossil fuel subsidy reform, even with a reduced proportion of nuclear energy generation, was likely to bring multiple benefits to OECD countries, primarily in the forms of reduced energy generation costs and low-cost climate change mitigation.

Nuclear power generation has historically provided a significant source of near zero-carbon emissions. If nuclear power continues to be phased out in the wake of the Fukushima disaster, it is likely that it will be replaced by fossil fuel generation sources, since they are often the lowest cost options in the short run. This reality necessitates strategic climate policy to direct the energy alternatives toward renewable sources. The removal of nuclear power from energy mixes would likely have negative consequences on carbon emissions unless offsetting policies, such as fossil fuel subsidy reforms, are enacted. The authors contend, however, that fossil fuel subsidy reforms are not likely to produce sufficient reductions from emissions and that the establishment of carbon pricing policies are likely critical for meeting international climate goals such as the Copenhagen Accord.

Conclusions

The advantages that nuclear power has to offer often make it an economically and environmentally feasible source of energy. The benefits of its low operating costs, functionality as a baseload power source, and low greenhouse gas emissions may often outweigh the costs of its high demands on water and land and the risks of nuclear reactor meltdowns. Nuclear electricity generation is viable as a baseload energy source in many scenarios and independent case studies should be conducted before increasing or decreasing capacity.

References Cited

Akashi, O., Hanaoka, T., Masui, T., Kainuma, M., 2014. Halving global GHG emissions by 2050 without depending on nuclear and CCS. Climatic Change 123, 611-622

Bird, D. K, Haynes, K., van den Honert, R., McAneney, J., Poortinga, W., 2014. Nuclear power in Australia: A comparative analysis of public opinion regarding climate change and the Fukushima disaster. Energy Policy 65, 644-653

Byers, E. A., Hall, J. W., Amezaga, J. M., 2014. Electricity generation and cooling water use: UK pathways to 2050. Global Environmental Change 25, 16-30

Magné, B., Chateau, J., Dellink, R., 2014. Global implications of join fossil fuel subsidy reform and nuclear phase-out: an economic analysis. Climatic Change 123, 677-690

Poinssot, C., Bourg, S., Ouvrier, N., Combernoux, N., Rostaing, C., Vargas-Gonzalez, M., Bruno, J., 2014. Assessment of the environmental footprint of nuclear energy systems. Comparison between closed and open fuel cycles. Energy 69, 199-211

Soria, B., Ruiz-Sánchez, R., Estadieu, M., Belda-Sánchez, B., 2015. Impact of the Taxes on Used Nuclear Fuel on the Fuel Cycle Economics in Spain. Energies 8, 1426-1439

Vine, D. & Juliani, T., 2014. Climate Solutions: The Role of Nuclear Power. Center for Climate and Energy Solutions

Woo, C. K., Ho, T., Zarnikau, J., Olson, A., Jones, R., Chait, M., Horowitz, I., Wang, J., 2014. Electricity-market price and nuclear power plant shutdown: Evidence from California. Energy Policy 73, 234-244.

Section II—Biology and Ecology

Biological Responses to Climate Change

Anna Alquitela

Global climate change has occurred numerous times throughout Earth's history. However, the current state of warming that has happened in the last 100 years, has transpired much more quickly than ever before because of human involvement (Qin *et al.* 2013). The burning of fossil fuels, deforestation, and industrialization are just a few of the human-induced sources of greenhouse gas emissions that have contributed to climate change. Greenhouse gases cause heat to be trapped in the lower levels of the atmosphere. The heat is affecting temperatures, rainfall, and sea levels, which in turn are affecting phenological and other biological responses of organisms. As these changes are occurring, biological life must adapt in order to survive.

This chapter demonstrates how some organisms are affected by climate change in differing ways. Observed biological responses to climate change range from reductions in organism size to shifts in migration patterns and predator-prey interactions. Some organisms, such as the Lyme disease-carrying tick *Ixodes ricinus,* have even adapted resilience to climate change that may prove itself to be detrimental to other life forms. Not all organisms have been required to adapt yet, but as the climate continues its pattern of change, humans can assume that eventually all life will also change.

Resilience of a Disease-Carrying Tick

Gilbert *et al.* (2014) have tested their hypothesis that as temperatures increase, so will the questing behavior of ticks (*Ixodes ricinus*) that carry the pathogen *Lyme borreliosis* that causes Lyme disease. Ticks have three life stages: larva, nymph, and adult. For a tick to advance to the next life stage, it must obtain a blood meal from a host—questing is the act of finding a host.

To test their hypothesis, Gilbert *et al.* collected nymph tick populations within the last two weeks of May 2012 from four sites in Europe: Aberdeenshire, Scotland; Denbighshire, Wales; Hampshire, England; and Auvergne, France. The nymphs were placed in nylon mesh tubes outfitted with wet sand and damp moss. There were 30 nymphs per tube and 10–12 tubes for each population. The tubes were then placed into an incubator set to 5°C for two hours to ensure that no ticks were questing before the experiment began. The incubator temperature was increased by 1°C every 24 hours. At each temperature, the number of nymphs questing was counted at five time-intervals over the 24-hour period. The maximum temperatures tested were 15°C for the UK nymphs and 14°C for the French nymphs. In their natural environments, UK ticks quest between 6 and 15°C and French ticks quest between 7°C and 14°C.

Using projected temperature changes from the Intergovernmental Panel on Climate Change, for 1980–1999 and 2080–2099, averaged over 21 models, Gilbert *et al.* calculated temperature change projections for springtime at the four tested sites. These data were used in a general linear mixed model, to compare temperature with the number of nymphs questing per population. The authors found a significant relationship between increased temperatures and increased proportion of ticks questing that differed between populations. They note that the highest proportion of questing ticks were from the northeast Scotland population and that the highest proportion of questing ticks occurred at the lower temperatures tested.

The results of the experiment show that ticks from cooler climates demonstrate a higher proportion of questing behaviors. Because their hypothesis has so far been supported by their experiment, the authors believe that the increased questing behavior of *I. ricinus* is indicating a potential resilience to climate change.

Reduced Body Size as a Positive Response

Surveying 85 unique sites on the western borders of North Carolina and Virginia, biologist Nicholas Caruso and his team collected data on adult specimens of Appalachian woodland salamanders (genus *Plethodon*). The results showed a reduction in salamander body size in accordance with lower climate temperatures. Because woodland salamanders are lungless, they breathe through their skin and require a moist environment for survival. The authors used historical and present-day data to model the changes in body size of 15 species of salamanders over the past 55 years. The dataset included 9,450 adult body size measurements from 102 populations of the 15 different species of salamanders. An 8% reduction in the average salamander size was found in all of the species over the 55-year study (Caruso *et al.* 2014). The reduction in body size reveals the plasticity of organisms to adapt to changes in climate. Because body size is directly linked to diet and foraging behavior, growth rates are also affected. Smaller body size means less surface area, and less loss of moisture through cutaneous responses. Also, the salamanders that were surveyed have demonstrated an increase in metabolism.

In a warming world, animals will need to adapt in order to survive. The woodland salamanders have revealed that, at lower latitudes of North America, where the forests have become drier over the last 55 years, they are at a low risk of extinction because of their ability to respond positively to climate change. The authors were not able to determine if the reduction in body size was due to changes in energetics caused by growth response, or if it was due to natural selection of a reduced size in conjunction with limited energy. However, the authors also speculate that higher temperatures may affect the developmental stage of salamanders, causing retardation in growth without negatively affecting adult survivorship.

Reduced Fecundity in Wood Frogs

From 2006 to 2012, Michael Bernard conducted research at a field station in southeastern Michigan where he used drift fences and pitfall traps to capture both adult and metamorphosing wood frogs (*Rana sylvatica*) at six wetlands (Bernard 2014). Bernard's goal was to determine if a relationship exists between date of breeding and winter temperature and precipitation, and between the female reproductive rate (fecundity) and winter temperature and precipitation. Using these data, he was also able to discern if breeding dates affect changes in metamorphosis timing, length of the larval period, weight at metamorphosis, and larval survival.

Bernard found that warmer winters led to reductions in female fecundity, possibly due to the increased energy consumption caused by warmer winters. Because wood frogs survive through winter by producing glycogen, a cryoprotectant that minimizes freezing damage to their biological tissues, Bernard states that there is a "metabolic cost" to thawing during warmer winters. During the time of his research, the difference in average maximum temperature between the coldest winter and warmest winter was 5°C. The number of eggs captured during this time was 15% of the average clutch size. Over time, this reduction in clutch size could negatively affect the population size of wood frogs. Bernard also found that when breeding occurred at an earlier date, the larval stage took longer, but the metamorphosis still took place earlier than when breeding occurred at a later date. He credits these effects to warmer weather causing tadpoles to develop more quickly; however, the developmental stage is still much slower. Bernard found no changes in larval survival due to shifts in breeding time, but did see an increase in average mass at metamorphosis during the course of his research.

Climate Change Helps to Prepare Geese for Migration

Greenland white-fronted geese, *Anser albifrons flavirostris*, spend their winters in Ireland, stage and refuel in Iceland, and breed in Greenland. Because climate change has advanced the spring thaw in

Ireland by about 18 days since 1985, these geese have more food to eat and are departing for and arriving in Iceland earlier than previous years. However, climate change has not caused a significant change in the temperatures of Icelandic staging areas and therefore has not caused a significant change in the departure time from Iceland to Greenland, thus the geese are staying in Iceland for a longer period of time than in previous years and arriving at their breeding grounds around the same time as historically known. Fox *et al.* (2014) used an abdominal profile index (API) as an indicator of fat stores in geese to determine if the amount of stored fat was the cause of advanced departure to Iceland. The authors also considered trends in temperature as indicators of departure time, but found that, because of the migration distances from Ireland to Iceland and Iceland to Greenland, temperatures are "very poor predictors" for departure times; meaning that the geese would not be able to use the temperature in Ireland to predict the temperature in Iceland nor the temperature in Iceland to predict the temperature in Greenland. Compared to previous studies, Fox *et al.* found that the Greenland white-fronted geese departed Ireland 33 days earlier in 2012 than they did in 1969 and arrived in Iceland 22 days earlier in 2012 than they did in 1997. The authors also note that the mean API at departure from Ireland in 2012 and 2013 increased significantly from previous years.

The study by Fox *et al.* has determined that, though temperature is not a determinant of migratory dates, climate change has affected spring thaw which allows for earlier grass growth, in turn leading to advanced fat storage in Greenland white-fronted geese and causing them to depart from their wintering grounds earlier than in previous years. The staging area in Iceland is at an agricultural university farm where there is more than enough available food for the geese. This is likely the reason why the geese are not migrating from Iceland to Greenland sooner. Because temperatures in Greenland have not affected snow levels there, it is important that the geese arrive in ideal health in order to reproduce optimally.

Ectotherms Produce a Lost Generation

Ectotherms are organisms that rely on environmental heat sources to maintain their body temperatures. The timing of reproduction in these organisms has much to do with their available thermal time windows. According to Van Dyck *et al.* (2015), the warming climate could cause an increase in the annual number of generations of ectotherms, but this response is potentially maladaptive, leading to a developmental trap; if an organism reproduces at the end of its reproductive season, there is a good chance that the new generation will die and the energy used to produce that generation would have been in vain.

Van Dyck *et al.* performed field experiments with the wall brown butterfly, *Lasiommata megera*, an ectotherm, to show that regional warming could be affecting their late summer suspended development. These butterflies always produce two generations per year, but during warm summers a third generation may also be produced. Van Dyck *et al.* introduced 253 young caterpillars into four Belgian sites, two inland and two coastal. They found that all of the caterpillars in the inland sites, where the native populations have disappeared, survived to the adult stage, whereas only 42.5% of the caterpillars in the coastal sites developed to the adult stage even though native populations persist there. During the experiment, both the mean ambient temperature and the mean temperature at host-plant height were recorded. The mean ambient temperature was 0.5°C higher, and the mean temperature at host-plant level, where the caterpillars are located, was 1.2°C higher at the inland sites than the coastal sites, respectively. The change in thermal conditions of the inland populations has caused a mismatch between seasonal cues and suspended development of *L. megera*.

Van Dyck *et al.* argue that the ectotherms that are most likely affected by climate change are those that "use photoperiod as an important cue for life-cycle regulation." They stress the importance of emphasizing cue-response systems in the field of ecology and phenological evolution to better understand the impacts of climate change.

Absorptivity Affects Wing Melanin in Butterflies

Climate change affects many organisms in varying ways. Organisms that are unlikely to migrate must adapt to climate change through evolutionary responses. Numerous studies have documented evolutionary responses to climate change over a period of one to three decades. However, the study by Kingsolver and Buckley (2015) provides evidence of a delayed evolutionary response to climate change for the subalpine and alpine butterfly, *Colias meadii*, in the southern Rocky Mountains of Colorado. Kingsolver and Buckley investigated the evolution of wing melanin of *C. meadii* and its effect on selection. They state that many species of the study area have adapted to their local climate conditions and "have limited potential for large-scale migration and gene flow." Comparing wing melanin to regional climate change, the authors relied on models to predict selection and evolutionary responses over the past 50–60 years. They also analyzed the changes in population mean fitness over time at the subalpine and alpine sites by relating solar absorptivity (the amount of radiation absorbed by the body) to fitness for a given site and year. Between 1955 and 2010, during the month of July, both the mean and daily maximum temperatures increased significantly for the study sites. Kingsolver and Buckley found that butterflies at the subalpine site had lower optimal absorptivity and a higher maximal fitness, whereas butterflies at the alpine site displayed a cohesive increase of optimal absorptivity and fitness, thus determining that the increased wing melanin exhibited in *C. meadii* has been favored at the alpine site. However, there are two opposing factors affecting fitness: increased melanin enables increased flight and reproduction, but it also increases thermal stress that negatively affects egg viability and adult survival rates. A premature evolutionary response to climate change would adversely affect the populations of *C. meadii*, therefore it would be expected that the butterflies that have survived to reproduce effectively are those that have a *slower* evolutionary response to climate change.

Nesting Behavior of Leatherback Turtles

Much research has been conducted on phenological reproductive responses to climate change. These responses occur commonly in plants, butterflies, birds, amphibians, fish, and insect larvae. Because sea turtles use thermal cues to begin migration to nesting sites, the authors hypothesized that leatherback turtles (*Dermochelys coriacea*) are delaying nesting in response to increased temperatures at their foraging grounds (Neeman *et al.* 2015). Biological responses due to increased temperatures have been observed in many other species of sea turtles. Some of these effects include offset sex ratios in embryos, nesting scarcity, reduced clutch size, and increased mortality rates of eggs and hatchlings.

Data were collected at the foraging and nesting sites of two beaches (Playa Grande and Tortuguero) in Costa Rica and one beach (Sandy Point) in St. Croix, US Virgin Islands over the course of nine seasons between 1983 and 2010. The data included observations of turtle tracks at nesting sites and average temperatures at foraging sites. Statistical analyses were performed for each nesting site, correlating the 5th and 10th percentile dates to the respective foraging site temperatures. Of 72 comparisons between nesting dates and temperatures, only 10 correlations were significant, and of the 20 correlations between nesting dates and seasonal timing of temperatures, only one was significant. The authors admit that "migratory cues are complex," but found that sea turtles are not responding to "changes in seasonality at the foraging grounds." It is feasible that global climate change is not affecting the reproductive success of leatherback turtles. Nevertheless, the IUCN Red List of Endangered Species lists leatherback turtles as vulnerable because they often mistake plastic bags that end up in the ocean for jellyfish, their favorite prey, and eating plastic bags is damaging or fatal to them.

Stresses of Climate Change and Predation Affect Foraging Behavior and Growth

Because predation risks already cause physiological stresses to prey, the added stress of climate change may be a tipping point—pushing the tolerance limits of prey populations over the edge. Miller *et al.* (2014) focus their experiment on the combined stress of warming and predation on the intermediate consumer *Nucella lapillus*, a snail found on the rocky shores of the North Atlantic coast.

The purpose of the paper was to observe the effects of climate change on the top-down interactions of the green crab, *Carcinus maenas,* which preys on the snail, *N. lapillus,* which feeds on the mussel, *Mytilus edulis.* For the 30-day experiment conducted during the summer of 2010, Miller *et al.* constructed 64 "mesocosms" (plastic tubs), each containing a perforated plastic chamber for a crab and above that, they placed a granite tile which was used as a base for a wire cage containing mussels and snails. The mesocosms allowed for the snails to forage on the mussels while in the presence of a predator, though without being at risk of predation. Within the mesocosms, the tides were simulated using drains and ball valves to control the water levels. To simulate high tide, the non-control mesocosms were filled with warmed seawater and, to simulate low tide, heat lamps were used to warm the upper chamber. Staying well below the mortality limits for *N. lapillus,* Miller *et al.* used submersible heaters to increase the water temperatures of the mesocosms by an average of 2.4°C and 3.2°C for high and low tide, respectively.

The authors found that the stresses of predation risk combined with warming temperatures caused the following effects on *N. lapillus*: a significant decrease in foraging behavior, a lack of tissue growth, and a negative growth efficiency. These changes in behavior pose a consequential effect on the intertidal natural community of *N. lapillus* and *C. maenas* because if the snails are not growing, their fitness will drop, eventually leading to a decrease in the population and causing a shift in predator-prey interactions.

Conclusions

As climate change continues to affect life on this planet, it is compelling to witness the adaptations that occur in response to that change. The goal of every species is to survive, and though meaningful evolutionary changes may take many generations to become evident, phenological and behavioral responses to a warming planet can transpire much more quickly. As seen in the research summaries included in this chapter, responses to climate change can have both positive and negative implications for life in any ecosystem. Niches can be filled just as they always have, and life on Earth will persist even as certain species go extinct. The important point that humans must understand is that we are the cause of this current climate change. Both our history and our ongoing decisions are affecting both human and non-human life in the present and the future. Each of us must become more considerate and thoughtful about the choices we make and their influences on all life.

References Cited

Bernard, M. F., 2014. Warmer winters reduce frog fecundity and shift breeding phenology, which consequently alters larval development and metamorphic timing. Global Change Biology. DOI: 10.1111/gcb.12720

Caruso, N. M., Sears, M. W., Adams, D. C., Lips, K. R., 2014. Widespread rapid reductions in body size of adult salamanders in response to climate change. Global Change Biology 20, 1751–1759.

Fox, A. D., Weegman, M., Bearhop, S., Hilton, G., Griffin, L., Stroud, D. A., Walsh, A., 2014. Climate change and contrasting plasticity in timing of a two-step migration episode of an Arctic-nesting avian herbivore. Current Zoology 60, 233–242.

Gilbert, L., Aungier, J., Tomkins, J., 2014. Climate of origin affects tick (*Ixodes ricinus*) host-seeking behavior in response to temperature: implications for resilience to climate change? Ecology and Evolution 4, 1186–1198.

Kingsolver, J. G., Buckley, L. B., 2015. Climate variability slows evolutionary responses of *Colias* butterflies to recent climate change. Proceedings of the Royal Society of London B: Biological Sciences 282, 20142470.

Miller, L. P., Matassa, C. M., Trussell, G. C., 2014. Climate change enhances the negative effects of predation risk on an intermediate consumer. Global Change Biology 20, 3834–3844.

Neeman, N., Robinson, N. J., Paladino, F. V., Spotila, J. R., O'Connor, M. P., 2015. Phenology shifts in leatherback turtles (*Dermochelys coriacea*) due to changes in sea surface temperature. Journal of Experimental Marine Biology and Ecology 462, 113–120.

Van Dyck, H., Bonte, D., Puls, R., Gotthard, K., Maes, D., 2015. The lost generation hypothesis: could climate change drive ectotherms into a developmental trap? Oikos 124, 54–61.

Qin, D., Plattner, G. K., Tignor, M., Allen, S. K., Boschung, J., Nauels, A., Midgley, P. M., 2014. Climate change 2013: The physical science basis. Cambridge, UK, and New York: Cambridge University Press.

Saving Endangered Species

Alexander Birk

The twenty first century is riddled with problems facing the natural world. Humans have put an increasing strain on the environment over the years. Anthropogenic effects on many aspects of nature are visible. Whether the damage is done by population increases, technology innovations, increased energy consumption, or increased need of natural resources. The effects that humans have are everywhere. A huge part of the natural environment affected is the wildlife.

There is a long list of endangered species in the world, and it continues to grow as ecosystems become more damaged. People have become more aware of things such as climate change, but have little idea how they affect individual endangered species. With increasing damage, there are more endangered species, along with more species extinction as well. In order to protect endangered species, people need to realize the effects of their actions.

While people are trying to solve the climate change issue, there are still consequences in the solutions that people have to climate change. When it comes to energy, the solution to reducing greenhouse gases may also have a negative effect on endangered species. For example some sustainable energy resources may destroy natural habitats, this makes the situation difficult to solve. There are bound to be sacrifices, but people need to make sure that they do not sacrifice by allowing more species to become endangered and extinct.

With more extinctions come a reaction bringing even more problems. Most ecosystems will have a significant negative effect on the entire system if they lose a species. Most people do not realize that sacrificing a single species may very well have severe effects on the entire ecosystem. If people do not take endangered species seriously, even more will become endangered, and the world as we know it will completely change.

The Endangered Species Act: Conservation-Reliant Species

The United States Endangered Species Act (ESA) is responsible for the protection of species and their habitats. The ESA maintains a list of the species at risk; the ultimate goal is to get these species off of the list. In order to get an endangered species removed (delisted) the ESA must regard that species as self-sustaining. The definition of a self-sustaining species becomes difficult as the ESA looks at conservation-reliant species. A conservation-reliant species is defined as a species that has been delisted; however it requires management in order to prevent it from once again being at risk.

There are multiple dimensions that qualify a species as conservation-reliant; the two categories are either one time intervention or continuous intervention. A few examples of one time interventions would be dam removal or habitat restoration. Continuous intervention would be defined by actions like translocation or invasive species control. Most species that have been classified as conservation-reliant species are ones that require some of the following conservation action: controlling other species, controlling pollutants, managing habitats, controlling exploitation by humans, or assistance in population growth. Per ESA mandate these species are still self-sustaining outside of requiring these actions, therefor justifying delisting these species.

This becomes problematic as the focus of restoration shifts away from these species. Many conservation-reliant species are in need of intervention due to human interaction such as pollution, as mentioned previously. Even though this pollution does not put the species in a high risk of immediate danger, the intervention is needed to prevent long-term harm. The popular view is to focus on populations of species in greater danger. However even though conservation-reliant species may also be classified as self-sustaining, this should not in turn lead to them needing less assistance. People generally decide against helping a cause that does not seem as though it is an immediate concern. This is a problem that the ESA is faced with as they choose how to label different populations of species.

Protecting Island Biodiversity

Island biodiversity is of paramount importance on a global scale. The islands of the world contain twenty percent of all terrestrial plant and animal species. In addition the rate of endemic species on islands is much greater than on main lands, and island species are facing many threats. Over half of the most recent extinctions on the planet come from species inhabiting islands. In addition one third of all terrestrial species that are currently threatened with extinction are island-dwelling species (Couchamp *et al.* 2014: Climate Change, Sea-level Rise, and Conservation: Keeping Island Biodiversity Afloat).

The threats to island biodiversity have not gone unnoticed, there are increasing island conservation efforts in place today. Most of these efforts consist of conservation of habitat, and also include invasive species removal, but the threat of sea level rise has not be as thoroughly considered, even though it is one of the most widely accepted effect of climate change. With the current rates of climate change sea levels are expected to rise anywhere from 0.26 to 2.3 meters by 2100, some islands will endure dramatic habitat change. Up to 19 percent of 4,500 island biodiversity hotspots could be completely submerged. This would mean that up to 300 endemic species would be threatened with extinction. There are currently 604 islands that have ongoing projects to help support endangered species survival. It is predicted that twenty six of those islands would be completely submerged if sea level rose one meter. This would not only inevitably wipe out a great number of island species, but it would also render all of the previous conservation efforts meaningless.

Islands becoming completely submerged is not the only issue as sea levels rise. The islands that do not become completely submerged will surely become smaller. This will push the native species closer together, which will in turn create more competition and possibly more extinctions. Areas inhabited by humans will be pushed inward as well, constricting even more space for native species. One option for protecting these island species would be translocation. However so many species are specifically adapted to their island environment it would be a challenge to find another place for them to thrive (not to mention the consequences of island species invading and disrupting other ecosystems). If the decision is not to move them, then the risk becomes losing even more island species to extinction.

Strategies for Preventing Climate Change Induced Extinction

Climate change is an increasingly problematic obstacle in slowing the rate of species extinction and has been widely accepted as the major threat to biodiversity. With this threat rising, there needs to be a better strategy to protect species. Identifying a species as in danger of extinction is the key to saving them, the current system to detect potential extinctions does not give enough time to effectively save some species once they have been deemed endangered. According to Stanton *et al.* (2014) it would be much more effective to put species into more specific categories leading up to extinction. This would allow for a more visible pattern of endangerment over the years leading to a trajectory of where the species is headed. Stanton *et al.* (2014)

The trajectories made by Stanton *et al.* have been formed for three climate change scenarios: no change in emissions, reduced carbon emissions, and high carbon emissions. The trajectories will span 36 unique habitats through the year 2100. When a species begins to meet certain criteria it is identified as a "Red List" species. As a species approaches the "Red List" it can be tracked more closely. With earlier awareness, the rates of survival are projected to increase dramatically. Ultimately the trajectories will simulate how species' population patterns can tell us about the possible trajectory of the population as a whole. The goal is to find out if this new form of categorizing species could help ease the rates of extinction.

The different scenarios affect the outcome due to the effects of carbon on the subject species' habitats. In each model the effects on the environment are critical in the projections of species. As carbon levels rise, habitats become depleted and extinction rates are projected to rise. The rate of red list species dramatically increases as the carbon levels are raised. With the new proposed trajectory model, the signs of endangerment are found earlier than with the current model. Being able to see trends in populations has a significant effect on the possibility of species survival. Having this ability can let people know how and where to focus conservation efforts.

The Role of Scavengers in Ecosystems

Vultures act as great scavengers for many different ecosystems. Scavengers are well known for preying on the dead carcasses in their habitat. Vultures, as described by Campbell, are the epitome of natural scavengers. Scavenging is a very difficult way of life and many animals cannot do it successfully. The vulture is the perfect fit for a very specific niche as a scavenger in their ecosystem. Even though the various vulture species are successful scavengers, the exposure to anthropogenic affects may prove to be detrimental to the species' survival. Campbell (2014)

Vultures prove to be valuable assets to their ecosystems. Consuming dead carcasses is primarily helpful because it decreases the amount of disease. A vulture's strong digestive system can handle rotting carcasses. If left alone, these carcasses may cause a widespread epidemic among a species. Vultures do not hunt, or eat live prey, therefor they dominate the scavenger niche in their ecosystem.

According to Campbell vulture populations have experienced significant anthropogenic affects. Prior to humans, many large herds of animals led to plentiful food supply for scavenging species. Today those plentiful herds have been replaced with things like road kill, medicated livestock, and overall less food for vultures. The common veterinary drug diclofenac, often used for keeping healthy livestock, has been found to be extremely deadly to vultures. In addition the chemical substances found on road kill have proven to be harmful to vultures as well.

Campbell refers to vultures as permanent scavengers and their presence in an ecosystem does not allow for other temporary scavenging species to thrive because the other scavenging species are not as fit for the niche as vultures. If vultures are not present, the ecosystem lacks the full benefits that the scavengers offer. With various species of vultures heading towards endangerment, the value of their presence is becoming more apparent. While humans may not see the species as valuable, they hold an irreplaceable role in their ecosystems.

Many vulture species around the world have become endangered. Mostly due to anthropogenic affects, either dying off from human-introduced chemicals or starving from humans removing their food source. Vultures are crucially important in many ecosystems, but most people think of them as dirty pests. Without vultures the biodiversity of many ecosystems may be in danger.

Renewable Energy and Endangered Species

As climate change continues to draw more attention, so does renewable energy. Creating completely clean energy to avoid emitting greenhouse gases seems like an unbeatable solution. However renewable energy projects often run into problems with the Endangered Species Act. Robbins (2014) claims that all renewable energy projects have one thing in common; they all destroy natural habitats and play a part in killing wildlife. Robbins (2014)

Climate change has numerous negative effects on many species, so it is common for advocates of climate change mitigation, and of biodiversity work together. However one of the major paths for climate mitigation, renewable energy, poses an even more immediate threat to biodiversity than climate change. Many renewable energy projects would take up vast amounts of land and render them uninhabitable for their existing species. The immediate destruction of habitat creates a greater concern for biodiversity advocates than the gradual degradation of habitat caused by climate change.

Biodiversity advocates have a powerful piece of legislature supporting their cause, the Endangered Species Act (ESA). Using the ESA, entire renewable energy projects can be shut down once an endangered species is found on site. This creates a problem because the three major renewable energy sources in the United States are solar power, wind power, and hydropower; all of which take up a great amount of space, and significantly alter their surroundings. This makes it very easy to find an endangered species that may be affected by the project.

In order to move forward with renewable energy projects one suggestion is that the ESA needs to prioritize what species are the most crucial to protect. This option suggests that not all of the species can be saved so it is better to build renewable energy projects on land where less important species are present. However biodiversity advocates argue that all species are important and it is impossible to determine which one would be more expendable. This brings up the next option, translocation. Moving species to another habitat similar to their native one could be helpful, but it could also be harmful to the habitat to which they are moved, potentially causing new species already there to become endangered.

The conflict between biodiversity and climate regulation is something that needs to be given more attention. We need to be able to find a way for these two ideas can work together so we can move forward. The end goal for biodiversity protection includes inputting climate mitigation strategies for long-term success. Thus it is reasonable to think that these two projects will be able to come together and create a solution that will make a better future.

Effects of Urbanization on Natural Habitats

Humans have always had a profound impact on natural environments, and McKinney (2015) suggests that one of the worst effects habitats endure is caused by urban sprawl. As human population increases, the need for more housing and other services increases. Development rates have shot up in recent years to fulfill the needs of the human population. Each piece of land that is developed causes a reaction in the natural environment; as habitat is removed, native species either die or need to relocate. This habitat loss can cause a significant loss in population of native species, and the ones that do survive are overcrowding a smaller habitat. In addition, a certain amount along the outside edge of the habitat is also affected by humans. This "edge effect" harms the native habitat along the edges of developed land so that it is also uninhabitable by native species. With that being said people are really disrupting more native habitat than they think when developing new land. The land that is actually developed has an impact on the environment as well. As you often see, animals are also living inside urban areas. The buildings and streets become part of the ecosystem as well. This is important to remember as we discuss conservation.

In order to avoid losing numerous natural habitats and native species, conservation efforts need to be utilized to their fullest. McKinney (2015) describes this as anything from avoiding building in certain areas, to making the developed areas less impactful. A great way to help ease the effects on the natural environment is to make the human-built environment less damaging. Doing this would not be too difficult for the average city planner to implement. Doing things such as keeping native plants as a part of the landscaping in the city can benefit native species tremendously by giving them places to live. In addition to planting native plants, having more plant coverage in cities is a crucial way to avoid the damaging effects of development. McKinney (2015) suggests that having more parks and more plants in the city could increase the species richness of the urban areas of the city. Ultimately if we try to lessen the damage of developed areas then we can successfully reduce the edge effect and lessen the pressure on numerous species. People are not going to stop developing, but they can start developing smarter by keeping the natural environment in their minds while planning.

Restoring Ecosystems After Climate Change

Raised awareness about climate change has caused an increase in action. Conde *et al.* (2015) support the fact that climate change is a key component affecting biodiversity across the globe. Even with the raised awareness of climate change, according to Conde *et al.* biodiversity has yet to see sufficient positive progress. This raises the question that if climate change is truly negatively affecting biodiversity, then why is it not having a positive turnaround as climate change awareness increases? A possible reason that Conde *et al.* suggest is that the level of reduction in climate damage has not reached a point to reverse the affects that it has already made.

Preventing species extinction is a major concern directly related to climate change. There are numerous studies showing the negative effects of climate change on biodiversity, but few that show if reversing climate change will help. In order to properly assist in the aid of ecosystems, it is important to recognize that the effects of climate change may need additional assistance. It is quite possible that we will need to look at certain environments case by case, and assess what each needs.

Many biodiversity hotspots are heavily affected by climate change. In most cases simply reversing the effects of climate change will not allow these hotspots to recover. These damages need human intervention in addition to minimizing climate damage. The habitat is not going to just completely resurrect itself once the effects of climate change are decreased. With human intervention, we can ensure that the biodiversity can make a steady increase as ecosystems recover.

Each ecosystem has been affected differently and thus should be treated differently in its recovery. Therefore conservation strategies are going to be key as we move forward in protecting biodiversity. While the efforts to reduce climate change are a crucial start, Conde *et al.* argue that more effort needs to be put into further restoring ecosystems. Full restoration of ecosystems across the planet will be a huge step in protecting biodiversity. Protecting and restoring the biodiversity of our planet is going to be a hard job, but with careful thought and planning it can be successfully done.

Effects of Endangerment on Trophic Levels

Most studies focus on endangered species at a singular level. Taking a case-by-case look at specific species is important, however looking at how one endangered species affects the entire food web is crucial. Species diversity is dependent on the entire ecosystem functioning together as a whole. Species loss at one trophic level can have dramatic effects on others as well. Barnes *et al.* (2014) look specifically at how trophic levels affect each other in tropical ecosystems. Topical ecosystems are perfect examples because the connection between species is extremely clear and relevant.

The example looked at further by Barnes *et al.* (2015) has to do with logging in tropical rainforests. With logging comes the possible destruction of habitats for many species. The species whose habitat is destroyed may have many effects on its surrounding species, and depending on the trophic level of the species, it may have different types of effects. The loss of habitat from logging may cause extinction, or severe endangerment. This can cause potential predators to lose out on main food sources and therefore cause a reduction and possible endangerment of those species as well. If the effects of logging directly affect a predator, or simply a species with a higher trophic level, this would have a different effect on the prey. The lower level species will not endure as much predation causing the population to vastly increase. That increase in population may cause harm in other levels of the food web. If it is a primary consumer then this may cause a raised pressure on some plant species. If it is an example in which a secondary consumer that experiences the raise in population then it will cause its prey to possibly become endangered.

Barnes *et al.* state that the tropical predators are at an especially high level of danger due to logging. With the loss of lower trophic species, due to loss of habitat, it has been observed that there is an increasing strain on higher predators. The energy loss in the system caused by logging directly effects the entire system. Therefore it is clear that logging is not only detrimental to the habitat of many species in tropical rainforests because of the loss of their habitats, but the entire system is damaged by harming a single species.

The balance operating in the natural world is a delicate one. Endangered species are increasingly affected by the damage of development and pollution, among other factors. People are finally becoming aware of the way they play a role in the environment around them, and beginning to take the right steps but there needs to be more of a sense of urgency. The endangered species count is rapidly increasing, and people are only beginning to understand the consequences of their actions. With new knowledge and new efforts, we can minimize the number of species that are lost.

References Cited

Barnes, A. D., Jochum, M., Mumme, S., Haneda, N. F., Farajallah, A., Widarto, T.R., Brose, U., 2014. Consequences of tropical land use for multitrophic biodiversity and ecosystem functioning. Nature Communications. doi: 10.1038/ncomms6351

Carroll, C., Rohlf, D. J., Li, Y.-W., Hartl, B., Phillips, M. K. and Noss, R. F. (2014), Connectivity conservation and endangered species recovery: a study in the challenges of defining conservation-reliant species. Conservation Letters. doi: 10.1111/conl.12102

Conde, A. D., Colchero, F., Guneralp, B., Gusset, M., Skolnik, B., Parr, M., Byers, O., Johnson, K., Young, G., Flesness, N., Possingham, H., 2015. Opportunities and costs for preventing vertebrate extinctions. Current Biology. doi: 10.1016/j.cub.2015.01.048

Franck Courchamp, Benjamin D. Hoffmann, James C. Russell, Camille Leclerc, Céline Bellard. 2014. Climate change, sea-level rise, and conservation: keeping island biodiversity afloat. Trends in Ecology & Evolution. doi:10.1016/j.tree.2014.01.001

Jessica C Stanton, Kevin T. Shoemaker, Richard G. Pearson, H. Resit Akcakaya. 2014. Warning times for species extinctions due to climate change. Global Change Biology. doi: 10.1111/gcb.12721

McKinney, L. M., 2015. Urbanization, biodiversity, and conservation. BioScience doi: 10.1641/0006-3568(2002)052[0883:UBAC]2.0.CO;2

Michael O Campbell. 2014. A fascinating example of convergent evolution: endangered vultures. Biodiversity & Endangered Species. doi: 10.4172/2332-2543.1000132

Robbins, K., 2014. Responsible, renewable, and redesigned: how the renewable energy movement can make peace with the endangered species act. Minnesota Journal of Law, Science & Technology, Forthcoming. 15:1, 555-583

.

Species Distribution Modelling and Climate Change

Kyle Jensen

The distribution of a species is determined by many interrelated factors, including the geology and effects of soil, the interaction of the species with other organisms, and climatic factors such as humidity, salinity, and temperature. The area of a range may match a certain soil type, the range of a prey species, or, as is the case for many species, a certain range of temperature-determined physiological limitations. This factor of temperature is particularly important today as we face a dramatic rise in global temperature due to anthropogenic global warming. The full effect of the resulting climate change will be extremely far-reaching, and many researchers have worked to give us some notion of what to expect moving forward. In this chapter I will focus on work that has been done to predict the future distributions of species as the earth warms, which requires some understanding of the current field of species distribution models (SDMs).

SDMs generally try to take into account the key factors which determine the range of an organism, and project for a given set of circumstances where that organism would be expected to appear. The geography is usually broken down into a grid, with calculations being done to determine the suitability of each square or pixel for the species in questions. It is necessary then that the researcher have some previous knowledge of what makes a suitable habitat for a given species, so as to generate the factors which will be put into the model. These factors are then adjusted until the projected distribution matches an already observed range where the intensity of the factors are known. So if there were four potential factors, these could be adjusted individually until the model best fit the empirical data from the field. Models that accomplish this are considered to have high fidelity, and are then considered relatively trustworthy in their projections of future distributions.

These future projections depend on having some sense of how certain variables will change over time. In these studies temperature is the main variable adjusted, and rather than spend the time to create their own climate models from scratch these researchers have taken advantage of existing data sets from sources such as the IPCC. When the model is ready they simply plug in the temperature data and let the model run its course to determine where exactly a species might be found in the future. And in these results we may find a great variability depending on the species in question.

It perhaps shouldn't come as a surprise that different species are expected to react in different ways to a warming climate. Those which are mobile are expected to move towards higher latitudes, a trend which is already being observed today for many species (Chen *et al*, 2011). The less mobile species will likely see contractions of their range, which is again being observed (Zhu *et al*, 2011). Others even are expected to see an increase in their range due to warming climates, the most notable of which are the many invasive species projected to move into areas which they had previously been unable to access (Dukes and Mooney, 1999).

Such changes will have profound environmental consequences, and the usefulness of these models lies in giving us a chance to figure out an idea of what exactly these changes will be like. This makes SDMs a powerful tool for conservation, as they can put to a wide variety of uses. One example is that knowing what habitat may become suitable for an invasive plant in the future allows us to determine where might be best to focus preventative efforts. Another is that if a species is projected to contract in its distribution, then we can use the SDM to determine which areas will remain suitable for the species and focus our preservation attempts there. The models allow us to determine how best to organize limited resources in mitigating the many and varied of climate change.

Now we are ready to take a look at some of the most recent efforts in this field. The topics range from the opening of the Northwest Passage to the extirpation of mountainous pika due to physiological limits. These studies exemplify good use of SDMs while seeking to further advance our understanding of how best to approach this research in the future. They are, I hope, a good representation of this fascinating and necessary field of study.

Are species distribution models validated by field trials?

Invasive species, especially plant species, are one of the greatest current threats to the Earth's biodiversity. It is feared that with the advent of global warming areas favorable to such species will increase, especially for those invasives from warmer climates that have naturalized near areas of marginal temperature. This could have negative impacts on the diversity of exposed populations, so species distribution models (SDMs) have been developed to estimate possible future distributions of organisms. These models make predictions by relating occurrence data to environmental conditions, giving a general idea of how the potential threat of an invasive species may change over time, and suggesting possible mitigation activities. Such models however have rarely been tested against experiments from the field. Sheppard *et al.* (2014) seek to validate SDMs through field trials at varying sites based on suitability as predicted by SDMs. If the predicted success of species in the models matches those of actual field trials, then we could be more confident in ability of models to assess the risk of invasive success. The experiment also addresses the validity of the enemy release hypothesis, which is often assumed to be the case in invasive studies. The hypothesis posits that invasive species leave behind any natural enemies when they are introduced to a new environment, which would contribute to their success. This experiment questions that assumption and its use in SDMs.

The experiment was conducted in New Zealand with three recently naturalized plant species chosen for observation: *Archontophoe-*

nix cunninghamiana, Psidium guajava and *Schefflera actinophylla.*
Each of these plants is a woody, bird-dispersed species, and thus likely
invaders, the perfect subjects for SDM. Three native plants were also
selected as controls to create an appropriate basis for comparison to
analyze the relative success of each invasive species. The alien palm *A.
cunninghamiana* for example was paired with the endemic *Rhopalostylis sapida* as both species are part of the same subtribe and live in
similar habitats. All six species were then planted at six field sites
across the two islands, the locations of which were chosen based on
predicted habitat suitability by SDMs. The sites were as similar as
possible in regards to soil quality and slope, to eliminate variables
aside from climate. Half of all plants were sprayed periodically with
pesticide to eliminate any natural enemies from both native and alien
species.

Over 18 months non-native species grew more than their native
counterparts in the 'suitable' areas but had a higher mortality rate in
the colder areas, though one site in particular saw more seed die-off
than expected. Thus the field results were mostly in line with the predictions of the SDMs, with discrepancies due possibly to the below-
average temperatures during winter or the variation of limiting resources across field sites. The models, now validated to some degree
by field studies, may give us a fairly accurate estimate of suitable
ranges for a given species, and thus likely changes in range due to
climate change. Based on this combination of field studies and
SDMs, the potentially suitable environments for these invasive species seem likely to increase as warming raises the minimum temperature. Yet one assumption of invasive studies generally may need to
undergo greater scrutiny as the enemy release hypothesis did not hold
in this trial, with both native and alien plants being subjected to similar levels of herbivory with or without pesticide. The current acceptance of this hypothesis may now be called into question, with
further testing required.

The Future Distribution of Harmful Algal Blooms

As the climate warms, algal blooms of certain harmful species are presenting an increasing threat to many biotic communities. Their introduction is often anthropogenic, and their occurrence is driven by eutrophication and changing climates. Their range may be influenced by rising temperatures, altered salinity due to runoff caused by climate change, and increasing nutrient loads due to increased development and fertilizer use containing favorable levels of nitrogen and phosphorous. Glibert *et al.* (2014) used a model incorporating climatic changes to predict the future distribution of these harmful algal blooms (HABs) under climate change scenarios. The study found a general increase in the distribution and presence of these HABs, though effects varied by region.

As these blooms constitute a global issue, the study looked at areas around the world including the Northwest European Shelf-Baltic Sea system, Northeast Asia, and Southeast Asia, with each system representing varying levels of development. The authors focused their efforts on two pelagic dinoflagellate species groups, *Prorocentrum* spp. and *Karenia* spp. These species have a global distribution, are generally the most harmful of the HABs, and have well-studies physiologies. *Prorocentrum* spp. generally have smaller cells, bloom rapidly in nearshore environments, and are destructive to food webs. *Karenia* spp. on the other hand are generally larger, bloom slowly in offshore environments, and are toxic with impacts on both human and wildlife health. These two groups are the most prevalent of the HABs, and are thus useful in projecting global changes of these systems.

The model used by this study is the Global Coastal Ocean Modelling System, which incorporated projections of sea surface temperature, sea surface salinity, and average nutrient concentration to predict the future distribution of HABs. Nutrient concentrations included the two ratios of $NH_4^+:NO_3^-$ and inorganic N : P, both of which influence the presence and success of algal blooms. The model was successful in replicating current conditions, indicating high fidelity. Projections showed variable outcomes by regions. In Northern Europe the number of months conducive to blooms was projected to expand, with a greater increase for *Prorocentrum* spp. than *Karenia* spp. In Northeast Asia we may see a small geographic increase of range, with a potential reduction of presence where the HABs are currently located. Southeast Asia might see little to no change for the range of *Prorocentrum* spp., with a contraction of the range of *Karenia* spp. These reductions may be due to the temperature range exceeding what is suitable for the growth of HABs.

These changes are due mostly to factors of temperature, as the model was built around changes caused by climatic factors. These projections indicate shifts in vulnerability of coastal systems to HABs as well as increased environmental impacts. This is made more important by the fact that the model likely underestimates potential effects, using conservative estimates of temperature while excluding anthropogenic sources of chemical adjustment so as to focus solely on the effects of climate change. Eutrophication is however projected to increase in future years due to increases in agriculture and industry, which could lead to increased toxicity and the further growth and spread of these algal species. The effects of these harmful algal blooms is thus quite likely to increase in the future; hopefully models such as this can help with managing this growing problem.

Arctic Warming and the Atlantic-Pacific Fish Interchange

For most of the Quaternary Period the inhospitable environment north of the Arctic Circle has served as a biotic barrier between Northern portions of the Atlantic and Pacific oceans. Through it is known that interchange across the Northwest and Northeast passages has occurred, currently only 135 of over 800 fish species found above 50° of latitude are found in both oceans. Continued warming may result in the reopening of these passages resulting an accelerated interchange of species between the Atlantic and Pacific as species follow favorable conditions into higher latitudes. This may also lead to increased movement of fishing and shipping vessels through these channels, which could facilitate further interchange. This has the potential to impact the food webs and biodiversity of systems in both of these oceans, the consequences of which would affect ecosystems currently comprising 39% of global marine fish landings. To analyze potential impacts of future species interchange, Wisz *et al.* (2015) has made forecasts of potential distributions for 515 fish species.

All species used in the study had been observed north of 50° latitude to a sufficient numerical degree. Niche-based models were used under climate change scenarios to simulate the progression of species through the passages over time. The models were implemented with occurrence data matched to monthly oceanographic condition records, such that the model fit the species' bioclimatic requirements. This type of analysis is well suited to studying species distribution shifts as the range of arctic species is strongly determined by measurable abiotic factors such as temperature and salinity. The interchange potential of given species was determined by that species predicted occupancy of a given area based on the availability and suitability of habitat within each passage. Potential impacts on community and trophic structures for a given species were evaluated by assessing the length and current trophic level of the species in question. The models were run to the year 2100 and divided geographic area into 'pixel' units, giving a 1, 10, or 50% predicted occupancy for a given pixel.

Results indicate likely shifts in diversity and restructuring of high-latitude ecosystems. By 2050 species will have begun to approach the passages, the majority approaching the Atlantic from the Pacific along the NE passage. By 2100 the number of potentially interchanging species will have increased sharply, as even more Pacific and some Atlantic species find more suitable conditions in the passages. Conservative estimates (based on 50% predicted occupancy) find that 13 species from the Pacific may reach the Atlantic, while 16% may reach the Pacific from the Atlantic, with almost all species moving across the NE passage. With more relaxed requirements (10% predicted occupancy), 44 species reach the Atlantic while 41 reach the Pacific, again mostly across the NE passage. Few species were considered likely to change the trophic structure or size of existing communities with the exception of the apex predators the Lingcod and Atlantic cod, which have the potential to dominate new ecosystems.

As of now, it is unclear what the final result of this interchange will be, though it will certainly continue well past 2100. The potential impacts as considered by this model are preliminary, and the authors call for further development in this area. Accounting for the interaction of species, as well as the mechanisms of spawning and dispersal, will allow for more accurate predictions of future species dispersal and community change. Such modeling will allow us to more accurately assess and analyze the potential impacts on biodiversity and environment future interchange may have on these hugely important populations.

Thermal Extremes and the Distribution of Species under Climate Change

Temperature is known to affect among other things the rate of population growth and survival of individuals, and an increasing number of studies are showing shifts in plant and animal distributions due to the direct influence of global warming. One method of modelling how future global warming will affect different species is through a trait-based approach, which tracks the effects of temperature events on a trait essential to fitness or distribution. Two popular model traits are the effect of temperature on organism growth rate, and the capacity of an organism to survive short periods of extreme conditions. Overgaard *et al.* (2014) compared these models to see whether the distribution of tropical and widespread *Drosophila* species would be more affected by average temperatures during growth or by the effect of thermal extremes on maintaining function. It was found that the ability of adults to tolerate thermal extremes was a better predictor of current species range than the thermal sensitivity of population growth.

Drospohila are ectotherms, meaning their body temperature closely matches that of their environment. Such species may be especially vulnerable to the effects of global warming. Ten *Drosophila* species were chosen, five being tropical specialists and five being widespread generalists. All had distribution records available for analysis. Populations of each were bred in captivity for controlled testing of thermal tolerance. Thermal limits of growth and development were determined by assessing egg to adult viability, developmental speed, and fecundity of all species in constant thermal environments at various temperatures, which were then compiled to form a measure of composite fitness reflecting optimal and critical limits of growth and development. Upper and lower thermal tolerance limits were determined by measuring at what temperature the flies were knocked into a state of coma. This was then compared against temperature data from 2002 to 2007 and estimates of future temperature in the year 2100.

Thermal tolerance limits were much better at predicting current distributional range than population growth, as most species experienced temperatures to just within their thermal limits. As an increase in the periods of extreme stress is expected under climate change, the *Drosophila*, at least, will become severely restricted in their range. Thus infrequent and extreme thermal conditions may impose critical limits to species distributions, which shows a harsher future for the distributions of species than models based on the effects of average temperatures alone.

Rising Temperatures and the Extirpation of Pika in California

The American pika, a cousin of rabbits and hares found in the mountains of western North America, may serve as a model organism for examining the effects of global warming on montane species. Pikas tend to live on talus slopes at higher elevations, and as their lower elevation limits are relatively high they may be especially vulnerable to climate change. Having adapted to colder climates pika are susceptible to hyperthermia in the summer, with a lethal upper body temperature occurring at only 3°C above their resting body temperature. During times of high temperature pikas reduce their foraging time to keep their body temperatures low, which also reduces their energy intake. Prolonged periods of high temperature can lead to reduced reproduction and death. Summer temperatures can thus place serious limits on the pikas' distribution. Stewart *et al.* (2015) created a model to assess the potential risk posed to pikas and other climate-sensitive mammals by climate change. Their model matched previous findings and predicted high levels of extirpation of pikas in study sites across California, with the size of talus area and summer temperatures being the best predictors of range.

The study was conducted across 67 sites in California with historical records of pikas, which were resurveyed to determine current pika presence. Pikas no longer occurred in 10 of these sites, with several other sites displaying only a small presence of the species. Pika viability was modelled as a function of environmental variables, resulting in 58 combinations which were tested for predictive power. The best performing model was based on the variables of talus area and mean temperature, which correctly hindcasted pika presence at 94% of sites. The model was then used to make predictions of future distribution under the increasing temperatures predicted by the IPCC. By 2070 overall extirpation of the sites is expected to range from 39% to 88%. The authors warn however that their model may underestimate extirpation due to the involvement of other environmental factors. Similar findings of range being linked with mean summer temperatures has been found across the pikas' range, which does not bode well for the species' future. And as the vulnerability to summer temperatures is shared by most high-elevation species, we face the risk of seeing these species 'pushed' off the tops of mountains by rising temperatures. The paper calls for greater monitoring of such species in the future.

Plasticity, Local Adaptation, and Climate Change Range Shifts

When assessing the future of biodiversity under such situations as climate change, it is almost always the species as a whole, which is used as a unit of measurement. Models of species distributions often assume that all individuals and groups within a species will respond with equal plasticity to environmental changes, yet this is often not the case. Research has shown that plasticity can be distributed unequally between individuals and populations, and that certain individuals and populations may also be more locally adapted to a given environment than others of the same species. Valladares *et al.* (2014) created a conceptual model of species distributions which incorporated phenotypic plasticity and local adaptation, and applied this model to both real experimental data and virtual species. They found that when differentiation between populations was accounted for and dispersal was restricted in the models, predictions of range loss under climate change became even larger.

In this conceptual model, populations within a species differed in the value of a trait related to fitness, producing various responses to climate change. Fitness was determined by the local adaptation (or genotype), the plasticity (or the translation of the genotype to the phenotype), and the strength of the selection upon a given phenotype. The model looked at five possible scenarios. The first was a scenario with no plasticity differentiation between populations, which is the implicit assumption of most species-distribution models currently in use, while the others each correlated with patterns of population plasticity observed in various studies of plants and animals.

The fitness-climate curves for 45 hypothetical subpopulations of 5 virtual populations were mapped across a European template using baseline climate data as well as future projections. Habitat suitability was projected for each of the populations according to the five different scenarios described earlier. When dispersal was allowed to be unlimited, geographic limits proved restrictive for northern populations regardless of plasticity, and southern and/or central populations (depending on the scenario) generally tended to do better than their northern counterparts, even expanding their ranges. When no dispersal was allowed, decline was seen for all populations in all scenarios, though range reduction was in some cases minimized due to plasticity. In applying the model to a real species, an Iberian tree *Pinus sylvestris* (Scots pine), they found that accounting for plasticity showed potential survival of southern populations under climate change scenarios.

The study shows then that plasticity can have a significant effect on the projections of ecological niche models. The literature review however showed contrasting patterns between species, so further study must be done before the exact effects of plasticity can be accurately determined.

Vulnerability of Trees and Biomes of the Rocky Mountains to Climate Change

As the reality of climate change becomes ever clearer, federal land managers are becoming increasingly concerned with how climate change will affect natural resources and ecosystem services within their jurisdictions. The western US in particular is expected to warm significantly, which will have widespread effects on the distribution of forests and various species; understanding these effects is essential to developing strategies to cope with future changes. Hansen and Phillips (2015) conducted a meta-analysis of five studies assessing the vulnerability of tree species and biome types to projected future climate changes, the results of which are expected to be used by the National Park Service as it initiates climate vulnerability assessments. The studies utilized bioclimatic envelope modeling, which showed a severe loss of territory for subalpine systems, especially for the white bark pine (*Pinus albicaulis*) and mountain hemlock (*Tsuga mertensiana*).

The geographic focus of the analysis included biomes within the Great Northern Landscape Conservation Cooperative (GNLCC) in the Northern Rockies as well as the ecosystems surrounding Glacier, Yellowstone, and Teton national parks. The tree species observed consisted of the dominant species within the study area, separated into three guilds; subalpine species occurring over 2400 meters, montane species which tolerate warmer and drier conditions and range from the lower tree line to mid elevations, and mesic species which enjoy warmer and wetter conditions. Vulnerability was based on four metrics of current and future climate suitability, including area of suitable habitat, loss of suitable habitat compared to the reference period, newly suitable habitat that will be naturally colonizable, and newly suitable habitat requiring assisted migration.

The rankings of vulnerable species were consistent among studies, showing that the subalpine guild as a whole was the most vulnerable, with all species projected to lose levels of habitat, while montane species were projected to expand in many areas. The expansion of suitability for mesic species was unclear. The two most vulnerable species were the western hemlock of the mesic guild, and the whitebark pine of the subalpine guild, which is projected to drop from 20% of observed area in the reference period to 0.5–0.7% by the end of the century. Such results should allow federal managers of these areas to prioritize more threatened species, and create strategies to adapt to our changing climate.

Effects of Alternative sets of Climate Predictors on Species Distribution Models and Estimates of Extinction Risk

As arid ecosystems have been recognized as being especially sensitive to climate change, they thus provide an appropriate system to assess the use of SDMs in estimating the threat of climate change to various species. Species distribution models (SDMs) can quantify relationships between species and environmental factors, and use this data to predict spatial distributions. SDMs are thus widely used to derive projections of species distribution under conditions of climate change. These models are correlative however, and as such are unable to identify causal species-environment relationships. They can only be used as supporting evidence for an existing hypothesis on factors affecting species distribution; as such the factors must be chosen as inputs for the SDM to function. Identifying the important climatic factors involved in determining the range of a given species is a key factor in assessing the potential effects of climate change on species distribution and extinction risk. Little research however has been done investigating the effects using alternative sets of climate predictor variables may have on the projections of SDMs. Pliscoff *et al.* (2014) seek to examine this area of potential uncertainty, addressing the potential variability of SDM spatial projections and determination of extinction risks through the creation and analysis of several sets of environmental predictors. They found that by adjusting climate predictor variables they were able to significantly affect predictions of spatial distribution as well as, for the first time, extinction risk estimates. This implies greater variability in such studies than previously thought.

The study modelled the present and future potential distributions of 13 species of the shrubby *Heliotropium* L. sect. *Cochranea*, a group with a range centered in the Atacama Desert of South America. They are well studied which means that their entire distribution can be recorded, making current and future projections possible. Presence data for these species dating back to 1950 were compared against several climatic factors, which were measured monthly over the course of a year. Six sets of variables were then developed, and eight modelling techniques were applied to each set of variables as well as to each species, resulting in 624 models. The effect of a set of variables on predictive performance and spatial projection were evaluated using generalized linear mixed models (GLMM). Climate change scenarios based on the six sets of variables were used to obtain future projections of distribution. Extinction risks were evaluated following the recommendations of IUCN, whereby predicted areas of occupancy (AOO) were incorporated into several GLMM analyses. Both range shifts and extinction risk were compared among the sets of climatic variables.

The use of different sets of climatic variables in SDMs significantly affected projections of spatial distribution, while having little effect on the predictive power of a given model. These uncertainties in distribution affect climate change projections, which in turn affect the IUCN estimates of extinction risk. This is the first time that the choice of climatic predictor variables has been shown to have impacts on extinction risk estimates. These newly recognized uncertainties should be further explored, and ought to be taken into account in future SDM studies.

Conclusions

Having reached the end of this chapter on species distribution modelling, I think it is important to acknowledge the limitations of these models. They are susceptible to error, and are almost certainly oversimplifying a complex situation, as was seen in some of these studies. Most models, for instance, don't take into account the relation of a species to other organisms, or the effects of decreasing population size, or the genetic isolation of subpopulations. Many studies, including several in this chapter, do seek to address these concerns and work to improve the function of SDMs. But such factors can be difficult to account for and, from what we can tell, knowing the general direction a species is going may be enough, especially if we temper such projections by closely observing the reality of the situation as it unfolds. There is a good hope that in using these SDMs we may be able to mitigate some of the negative effects on biodiversity wrought by climate change. And moving forward we will only get better at doing so.

References Cited

Chen, I. C., Hill, J. K., Ohlemüller, R., Roy, D. B., & Thomas, C. D. 2011. Rapid range shifts of species associated with high levels of climate warming. Science, 333, 1024-1026.

Dukes, J. S., & Mooney, H. A. 1999. Does global change increase the success of biological invaders?. Trends in Ecology & Evolution, 14, 135-139.

Glibert, P. M., Icarus Allen, J., Artioli, Y., Beusen, A., *et al.* , 2014. Vulnerability of coastal ecosystems to changes in harmful algal bloom distribution in response to climate change: projections based on model analysis. Global change biology, 20, 3845-3858.

Hansen, A. J., and Phillips, L. B., 2015. Which tree species and biome types are most vulnerable to climate change in the US Northern Rocky Mountains?. Forest Ecology and Management, 338, 68-83.

Overgaard, J., Kearney, M. R., and Hoffmann, A. A. 2014. Sensitivity to thermal extremes in Australian Drosophila implies similar impacts of climate change on the distribution of widespread and tropical species. Global change biology, 20, 1738-1750.

Pliscoff, P., Luebert, F., Hilger, H. H., & Guisan, A. 2014. Effects of alternative sets of climatic predictors on species distribution models and associated estimates of extinction risk: A test with plants in an arid environment. Ecological Modelling, 288, 166-177.

Stewart, J. A., Perrine, J. D., Nichols, L. B., Thorne, *et al.*, 2015. Revisiting the past to foretell the future: summer temperature and habitat area predict pika extirpations in California. Journal of Biogeography, 10.1111/jbi.12466.

Sheppard, C. S., Burns, B. R., & Stanley, M. C., 2014. Predicting plant invasions under climate change: are species distribution models validated by field trials?. Global change biology, 20, 2800-2814.

Valladares, F., Matesanz, S., Guilhaumon, F., Araújo, M. B., *et al.*, 2014. The effects of phenotypic plasticity and local adaptation on forecasts of species range shifts under climate change. Ecology letters, 17, 1351-1364.

Wisz, M. S., Broennimann, O., Grønkjær, P., Møller, P. R., *et al.*, 2015. Arctic warming will promote Atlantic-Pacific fish interchange. Nature Climate Change. 10.1038/nclimate2500

Zhu, K., Woodall, C. W., & Clark, J. S. 2012. Failure to migrate: lack of tree range expansion in response to climate change. Global Change Biology, 18, 1042-1052.

Responses of Corals to Stress from Global Warming

Kimberly Coombs

There have been five mass extinctions starting in the late Ordovician early Silurian period and going to the Cretaceious-Tertiary time period. Each of these mass extinctions has caused a loss of 50% to 85% of species present at that time and each is attributed to sea level changes and temperature changes, amongst other factors (Sheehan 2001; Erwin 1998; Hallam and Wignall 1999; McGhee Jr. 2001; Twitchett 1999, Erwin 1994; Marzoli *et al.* 2004; Jablonski and Raup 1995; Raup 1986). All of these mass extinctions appear to show the pattern that species are continuously vulnerable to climate changes in their environment.

Thus, climate change is an important factor to consider when studying the survivability of species, especially in today's world. Marine organisms, in particular, have shown declines in populations because of climate change; therefore, being able to predict the affect climate change will have on the density of species in the marine environment is of the utmost importance in order to ensure that species populations remain and ecosystems do not collapse (Greenstein and Pandolfi 2008). Global warming, an aspect of climate change, is one of the main factors that is negatively impacting the world's oceans. Global warming is a rise in the earth's temperature, which includes the temperature of the ocean's water. The temperature of the ocean is expected to rise by 1 to 3°C by 2100 (Great Barrier Reef Marine Park Authority 2009; Marshall and Johnson 2007; Munday *et al.* 2007).

Coral reefs are appearing as the most susceptible marine organisms to global warming, which, along with increased marine CO_2 concentrations and the resulting ocean acidification as well as increases in sea level, is having drastic effects on coral reef ecosystems (Marshall and Johnson 2007). For example, the Great Barrier Reef in Australia, the largest coral reef ecosystem in the world, has experienced two mass coral bleaching events in 1998 and 2002 (Berkelmans *et al.* 2004; Marshall and Johnson 2007), both of which were caused by global warming and resulted in a significant loss of corals and various fish species reliant upon these reefs (Munday *et al.* 2007; Marshall and Johnson 2007).

There are a couple of factors that cause coral reefs to be susceptible to increases in sea temperatures. First, part of the persistence of a coral reef is due to its size and whether it is well managed. For example, the Great Barrier Reef spans 344,400 km² (Great Barrier Reef Marine Park Authority 2009) and is well managed because it is a protected area. This means that the Great Barrier Reef will be able to withstand the consequences from global warming for a longer period of time than other smaller coral reefs. Second, coral reef ecosystems tend to be interdependent. This means that if corals become depleted in a reef, it may cause a ripple effect throughout the local ecosystem negatively affecting other organisms (Marshall and Johnson 2007).

Corals are highly susceptible to changes in sea temperature and play a critical part for the perseverance of coral reef ecosystems (Marshall and Johnson 2007; Great Barrier Reef Marine Park Authority 2009). The main source of calcium carbonate accumulation originates from corals, and they provide the foundation upon which the ecosystem is built (Marshall and Baird 2000). Corals not only build the foundation for the coral reef ecosystem, they also provide food as well as hiding places and homes for multiple species (Marshall and Johnson 2007). If corals were significantly reduced, the ecological impact would result in decreases in reef growth, species and habitat diversity, shelter for susceptible species, and an overall change in community structure (Marshall and Baird 2000). Furthermore, corals are known for living near their upper thermal limit; therefore, short periods of increased water temperatures put corals under a high amount of stress. The affects from stress on these corals may continue for 4 weeks even after the thermal stress is removed (Berkelmans *et al.* 2004; Berkelmans and Willis 1999).

Global warming's negative impact on corals is observable as what is currently known as bleaching. Bleaching is the phenomenon of corals loosing their zooxanthellae, which causes them to turn white or pale and can cause a decrease in growth and reproduction as well as an increase in coral mortality. It is predicted that as sea temperatures continue to rise, coral bleaching events will become more frequent and result in decreases in coral cover (Munday *et al.* 2007). There have been several recent studies that are trying to predict the fate of corals under a warming ocean.

This chapter focuses on various observations and model predictions of corals' survivability as sea temperatures continue to rise. Many of the model predictions are based on a business as usual scenario, meaning that no mitigation strategies are put in place to help prevent a loss of corals around the world. Some predictions and observations are more optimistic than others. One study in particular, concludes that corals will move northward and find new suitable habitats to populate in higher latitudes. Another study demonstrated that bleaching events cause coral cover to decrease and become dominated by macroalgae. Whether the various observations and model predictions are optimistic or not, the current state of corals will not remain the same under a warming climate.

Reef State and Resilience in a Climatically Changing Environment

It has been reported that the effects of greenhouse gas emissions have reduced coral reefs resilience, causing them to be more susceptible to stressors in their environment. As a result, coral reef state, the percent of coral cover, has begun to be greatly lessened, with a noticeable shift from coral dominated environments to macroalgae environments.

Bozec and Mumby (2015) conducted a study to look at the Caribbean coral reefs' resilience to increases in sea surface temperatures. There are acute and chronic stressors that arise from sea surface temperature increases. Acute stresses may cause unexpected mortality amongst coral reefs. Chronic stresses may influence the carbonate skeleton rate of extension, which impacts the mortality rates of size-dependent corals, rates of recovery among populations, and ecological interactions of coral reefs. Bozec and Mumby looked specifically at whether or not acute and chronic stresses from sea surface temperature rises would interact antagonistically, synergistically, or additively.

In order to evaluate the impact of rising sea surface temperatures, Bozec and Mumby developed a model of coral populations that could simulate the impacts of chronic and acute stresses. The model allows for chronic and acute stresses to be assessed together and separately. The model is also individual-based and has taken a mid-depth Caribbean coral reef as its model. The outputs of the model include the percent cover of the algal and coral species (reef state) from 2010 to 2060. Resilience of the coral reef was quantified every ten years from 2010 to 2060 as the probability that the coral reef was on a recovery trajectory. The thermal stresses encompassed in the model followed the RCP8.5 greenhouse gas emissions trajectory, which is simply the current greenhouse gas emission trends today and assumes there will be no mitigation strategies put in place.

Bozec and Mumby found that under no thermal stresses from increased sea surface temperatures, coral cover is able to proliferate over time. However, acute stress causes the reef state to show a significant reduction in coral cover by 2025 and chronic stress causes the reef state to show a significant reduction in coral cover by 2035. In terms of reef resilience, acute stress reduces coral reef's resilience and by 2060, coral reefs have less than a 50% chance of being able to recover. Chronic stress has less of an impact, as coral reefs have at least a 70% chance of recovering by 2060. When chronic stress and acute stress both impact the coral reefs, their affect is additive on reef state, and their interactions show synergism on coral reef resilience. Together, they cause a far greater affect on coral reefs with recovery becoming almost impossible by the year 2040.

The authors attribute these results to the reactions of corals to global warming. The reef state of corals is extremely sensitive to bleaching as bleaching results in loss of medium-sized to large corals that tend to have the greatest influence in coral cover. Bozec and Mumby believe that the two stressors are additive because of the way they act on the corals. Chronic stress acts directly on growth rate, reducing it with repeated bleaching events, and acute stress acts directly on mortality of corals; therefore, the interaction between these two stressors is indirect. For resilience, the acute and chronic stress interaction is synergistic due to both stressors working together to drive the coral reef towards the point of not being able to recover. Resilience and reef state are thus important indicators of a coral reef's health.

Potential for Coral Reefs to Recover after Coral Bleaching Events

The 1998 mass coral bleaching event caused much of the coral cover to be greatly reduced as many corals have a narrow set of temperature ranges that they can survive, and most live near their upper thermal maximum; therefore, slight increases in temperatures can have drastic negative affects on coral survivorship. Not much is known about the ability of corals to recover after coral bleaching events or the likelihood of the environment switching to an algae dominated environment.

Graham *et al.* (2015) conducted a study in order to identify reef recovery, the amount of coral cover being greater than macroalgal cover post-disturbance, or a regime shift, the amount of macroalgal cover being greater than coral cover post-disturbance, at the Seychelles reefs. This study observed 21 reef sites from 1994 to 2011 in which about 90% of the coral cover was lost in 1998. They found that 12 of the 21 reef sites were able to recover post-disturbance, yet it took about 10 years to see any major improvements in the amount of coral cover. On the other hand, the other 9 reef sites switched to a macroalgae dominated environment. Before the mass bleaching event, the macroalgae and coral cover percent were the same between the 12 reefs and 9 reefs, suggesting that the regime shift resulted from coral bleaching.

In order to evaluate and predict whether a reef recovery or a regime shift would result following a coral bleaching event, Graham *et al.* (2015) identified five specific factors that are good indicators of whether or not a reef will recover or undergo a regime shift: water depth, density of juvenile corals, nutrient conditions of the reef, reef's initial structural complexity, and the herbivorous fish biomass. Their results showed that reef recovery has a greater chance of occurring when the density of juvenile corals is more than 6.2 m^2, the reef is at least moderately structurally complex and at least 6.6 m deep, herbivorous fish biomass is low, and carbon:nitrogen ratios are high. Because it can be difficult to gather data on juvenile coral densities, nutrient levels, and herbivorous fishes biomass, the authors focused their attention on water depth and coral reef structural complexity, and were able to correctly predict whether a reef would recover or would undergo a regime shift 98% of the time, using only these predictors. Graham *et al.* (2015) also gathered data from six other countries that overall, these two predictors correctly recognize the trajectory a reef will endure after a bleaching episode.

Potential Outlooks on Coral Reef Structure as a Result of Climate Change

Coral reefs vary in structural architecture, meaning that the structure can be very complex or relatively simple. The more structurally complex a coral reef is, the more species diversity may be supported. The reef building corals that create the complex coral reef structures need to have a sustainable carbonate budget in order to continue the processes of accretion and erosion to build the coral reefs. These corals have been experiencing reductions in their carbonate budget; as a result, they have declined around the world.

Bozec *et al.* (2015) developed a mechanistic model that estimated reef topographical complexity under different scenarios for Caribbean reefs. They ran the model in order to investigate how Caribbean coral reefs topographical complexity may be impacted by global warming conditions in addition with hurricanes and management actions related to parrotfish population levels. They also ran the model to make predictions about the future of Caribbean coral reefs under the previously mentioned conditions. The model is able to calculate coral carbonate accretion and erosion on a coral colony scale. The model further simulates possible variations in coral surface and volume and the influence this may have on the topographical complexity.

Bozec *et al.* found that parrotfish populations may help prevent or lessen some of the negative impacts from global warming and hurricanes. Parrotfish aid in bioerosion on the coral reefs; therefore, keeping parrotfish grazing at relatively high levels will enable coral reefs to recover better from bleaching and hurricane disturbances. They also found that disturbances to the coral cover came more from hurricane disturbances than global warming, which they suggested is due to the strong currents that may lessen or prevent temperature changes occurring at these coral reefs. Bozec *et al.* believe that corals with strong skeleton structures form Caribbean coral reefs in order to withstand the impacts from hurricanes. Other studies have reported most losses in Caribbean corals due to hurricanes occurred within species that had delicate skeleton structures. Bozec *et al.* also noted the Caribbean corals reefs' resilience to global warming and hurricanes is due to the corals' ability to recover relatively quickly to these disturbances. The corals accretion rate outweighed the erosion of their carbonate budget from such disturbances; therefore, allowing the coral reefs to withstand severe damages. As for future predictions, the authors' estimate that under current trends, the structure of the coral reefs may decay over the next three decades, while under high hurricane occurrences, the coral reefs may decay in less than twenty years.

There was an assumption the authors made when running their model that may interfere with the credibility of their predictions. They assumed that the dead and live coral skeletons mostly influence the coral reef surface. Thus, the simulation did not take into account the carbonate fragments and sediments that satisfy the spaces created between the skeletons of the corals. Bozec *et al.* further noted that their model may be slightly simplistic; however, it does take into account information that is currently known, where as a more complex model would be less reliable due to the sheer knowledge gap. Nevertheless, Bozec *et al.* suggest that local management should aim to sustain high levels of parrotfish grazing in order to maintain Caribbean coral reefs topographical complexity under the current rate of disturbances and trends.

Algal Symbionts may make Corals Resistant to Rising Sea Temperatures

Corals share a mutualistic relationship with algal symbionts, but with increasing sea temperatures, these symbionts become expelled from the coral. The loss of symbionts causes the corals to become bleached and there have been declines of coral cover worldwide. Recent research has shown that there may be symbionts that are thermotolerant, such as the genus *Symbiodinium*, which may help reduce the amount of bleaching episodes seen amongst corals. *Symbiodinium* is divided into nine subgeneric clades, A-I, and the *Symbiodinium* D1a has been documented showing thermotolerance.

Silverstein *et al.* (2015) explored the impacts of increased sea temperatures of corals that contain different *Symbiodinium* clades. They collected colonies of *Montastaea cavernosa*, which they exposed to either a heat stress of 32°C or herbicide DCMU at 24°C for ten days. After ten days, the corals entered a recovery period of three months before half of them were exposed to the same stressor for another ten days and then placed back into a three-month recovery. Before the corals underwent their bleaching treatments, they contained *Symbiodinium* clade C. In both treatments, corals lost the same proportion of symbionts, and corals with a higher abundance of symbionts before treatments lost a larger proportion of them. *Symbiodinium* clade D was discovered to dominate the type of symbionts observed in corals after the three-month recovery period.

When corals were exposed to a second bleaching treatment, Silverstein *et al.* found bleached corals that developed *Symbiodinium* clade D1a lost fewer symbionts than corals still only containing *Symbiodinium* clade C from the first treatment. Between the two exposures (heat stress and herbicide DCMU), the amount of *Symbiondinium* clade D1a lost was the same amongst the corals; however, corals that had a higher abundance of clade D1a symbionts lost a greater proportion of them. Corals who also contained more clade D1a symbionts lost less of their symbionts than corals who contained clade C symbionts. During the second recovery phase, *Symbiondinium* clade D dominated the symbionts observed in the corals.

Corals that remained unbleached were found to contain low amounts of *Symbiodinium* clade D1a, which caused them to experience more acute bleaching when exposed to heat stresses. Silverstein *et al.* discovered that corals acclimated to 29°C also did not gain any thermoteralance when exposed to a higher temperature of 32°C. Thus, the corals that bleached in the first phase gained an advantage during the second phase of the bleaching treatment. Bleached corals gained a higher amount of *Symbiodinium* clade D1a that helped them be more resistant to bleaching and loss of symbionts in the second phase of bleaching.

Silverstein *et al.* observed that corals that had developed clade D symbionts and did not undergo the second phase of bleaching treatment, reverted back to clade C symbionts. *Symbiodinium* clade D may be thermotolerant, yet they cause the corals to grow more slowly, experience lower fecundity, and be less effective in translocating sugars.

Therefore, clade D symbionts may be useful in aiding the corals under high heat conditions, but when conditions are less stressful, clade C symbionts are more efficient for coral function.

Symbionts Impact the Behavior of Coral Larvae

Climate change is known for causing adults corals to become bleached, but it is also affecting the early life stages of corals. The larval stage of a coral reef's lifecycle is very important to its survivorship.

Corals disperse their larvae out into the water, then the larvae are responsible for finding a suitable substrate to settle on. After settlement, corals are able to start growing accumulate symbionts. Several studies have observed how different symbionts influence juvenile coral growth rates and thermotolerance; however, no data currently show if there are any influences from symbionts on the coral larvae before settlement occurs.

Winkler *et al.* (2015) conducted a study to observe how symbionts may affect coral larvae habitat choice, survival, and settlement success. They looked at *Acropora millepora* from the Great Barrier Reef, a coral species that usually associates with sensitive or thermally tolerant symbionts, and exposed the larvae of this species to four different temperature treatments ranging from 22.5°C to 28.5°C. The larvae were also exposed to seven different symbiont treatments that involved three different species of *Symbiodinium*: C3 (usually found in adults of this species), C15 (thermotolerant and not normally associated with this species), and D1 (thermotolerant and found in this species in certain regions). Habitat choice was documented based on where the larvae settled (on the top, back, or vertical surfaces of tile and on crustose coralline algae).

Lower temperatures were found to negatively impact *A. millepora* habitat settlement orientation and recruitment success, while other studies have found that higher temperatures negatively affect larvae success. Winkler *et al.* attribute their findings to their highest temperature being set at 28.5°C, where as other studies have had their highest temperature at 30°C or 32°C . This temperature was used in preference to 30°C or 32°C in order to represent a 3°C increase from the seasonal average temperature at the Great Barrier Reef, as only a 1–2°C increase in temperature is expected to occur at the Great Barrier Reef by 2100.

When temperatures were favorable (higher than 24.5°C) coral larvae settled on vertical surfaces. Vertical surfaces in the natural environment represent areas where low algal growth levels are maintained and there is no destructive grazing on the corals. Coral larvae also settled on crustose coralline algae in favorable temperatures. Crustose coralline algae represents a habitat that is suitable for settlement and later metamorphosis for coral larvae. Lower temperatures caused the coral larvae to choose suboptimal substrate selection (coral larvae chose to not settle on crustose coralline algae nor on vertical surfaces), therefore, this reduces the recruitment success.

The different symbionts that the larvae were exposed to showed different temperature-dependent crustose coralline algae selection. At a temperature greater than 26°C, larvae offered the symbiont D1 and C3 had a similar crustose coralline algae settlement selection, yet at lower temperatures, larvae offered symbiont D1 chose crustose coralline algae as their settlement more than larvae with symbionts C15 and C3. Larvae given the symbiont C15 chose to settle least on crustose coralline algae across the different temperatures.

The results that Winkler *et al.* found suggest a link between symbionts and settlement on crustose coralline algae. Crustose coralline algae metabolism is reduced at lower temperatures; as a result, this reduces the metabolites that attract symbionts and decreases the number of symbionts settling on crustose coralline algae. The results indicate that either different symbionts influence the crustose coralline algae settlement cue or that settlement is influenced by different symbionts behaving as a co-inducer.

Coral Bleaching Mitigated by Large-Amplitude Internal Waves

Coral reefs may be able to persist under a changing climate through multiple elements, such as thermotolerant symbionts or through extrinsic factors (environmental conditions) and intrinsic factors (community assemblages). For example, thermal stress on corals ranging in depth may be alleviated by large-amplitude internal waves that carry with them cold subpycnoline water.

One area in particular where large-amplitude internal waves affect coral reefs is the Andaman Sea. Wall *et al.* (2015) conducted a study to observe whether or not large-amplitude internal waves could offer some sort of environmental resistance to coral bleaching under thermally stressful conditions. They had seven sites exposed to the large-amplitude internal waves and five sites that were sheltered from these waves. Temperature and coral reef images were taken throughout the study period. Several different regression models and a community bleaching susceptibility index were used to analyze the data.

Wall *et al.* found that the large-amplitude internal waves tended to coincide with warmer sea surface temperatures thereby cooling the coral reefs. The amount the water temperature dropped varied across each of the exposed sites with the largest temperature drop reaching 22.1°C from a high of 31.9°C . The sheltered sites did not receive any cooling and were fully exposed to any thermal stress present. As a result, the percentage of bleached and dead corals was higher in the sheltered sites compared to the sites that were exposed to the large-amplitude internal waves. Though large-amplitude internal waves did not completely prevent bleaching from occurring, they did cause milder bleaching events, even in the species of corals that are most susceptible to bleaching.

There are several aspects of large-amplitude internal waves besides the cold water that may help corals be more resistance to bleaching events. First, they bring plankton and other nutrients to the corals. This allows corals to have more energy stores, which they can use to better cope with thermal stresses. Large-amplitude internal waves also bring about sedimentation and turbidity to the shallower coral reef regions. By doing so, light levels are reduced, which can ameliorate a bleaching event. However, removing the sediment is energetically expensive for corals and turbid waters also decrease coral photosynthesis, which Wall *et al.* believe is why they did not observe any differences in coral recovery between exposed and sheltered sites. Though corals recovery process might be slowed, the decreased degree of bleaching intensity from large-amplitude internal waves may outweigh this extra energetic cost if it means corals are better able to persist under high thermal stress conditions.

Coral Reef Biodiversity under Threat from Rising Temperature

Increases in sea surface temperatures are threatening the persistence of coral reefs' biodiversity and ecosystem services. The negative affects on coral reef ecosystems may not be as ominous as many models have predicted. About 125,000 years ago, during the Eocene Epoch, the world's sea surface temperature was about 6 °C warmer than today and coral reefs were found to populate in the mid-latitude waters. Though some decline in coral reefs occurred during this time, this period does give hope to coral reefs' survival under increasing thermal stress.

Descombes *et al.* (2015) used niche based models to assess the coral reef thermal suitability in the future and predict the most vulnerable locations to climate change based on Eocene and current coral reef distribution data. They looked at three different time periods, 2005–2014, 2050–2060, 2090–2100, and mapped the current and future sea surface temperatures with global ocean simulations and two different climate models. They further assessed whether or not some fish species, based on three categories, might be more vulnerable to climate change than others.

The model projects that the areas with the most species of corals and fishes will be the areas that suffer the largest biological decreases under rising sea surface temperatures. Thermally suitable habitats for coral reefs are greatly reduced when temperatures reach 32°C or higher. Tropical regions that support a high amount of coral reef diversity, such as the Indo-Australian Archipelago, is predicted to reach a temperature of 34.7°C by 2100; therefore, these regions will be the first to show a reduction in suitable habitat under a warming climate.

Descombes *et al.* indicate that in order for corals to find thermally suitable conditions, corals will need to move to higher latitudes. Corals in Florida and Japan have already started to shift towards higher latitudes. The problem with corals shifting northward is that it is unknown how much time it will take the corals to fully establish a suitable habitat in order to meet the demands of different fish species and if the different fish species can shift northward in time to escape potential extinction as their habitat is lost at the lower latitudes.

There are some limiting factors to this model though. For example, Descombes *et al.* assumed that fossil record data from the Eocene period they used in their model was based on thermally unfavorable conditions, meaning that corals not found during this time period resulted from an environment with temperatures that caused them not to persist. However, the authors note that it is likely that the fossil record did not fully represent the coral distribution. Descombes *et al.* also only took into consideration sea surface temperatures impacting corals when there are other factors that affect the survival of coral reefs, such as ocean acidification, sea-level rise, and salinity levels.

Nevertheless, Descombes *et al.* show that with increases in sea surface temperatures, coral reefs will shift towards higher latitudes. Coral reefs may become populated by other species as this shift occurs and the ecosystem services coral reefs provide will change and may be greatly reduced.

Economic Value of Coral Reefs

Coral reefs are known for supporting a habitat rich in species diversity and abundance. Besides the benefit coral reefs provide to other species, they also offer a benefit to humans. Coral reefs provide a source of economic gain in terms of tourism and fisheries, usually bringing in about $30 billion each year. However, climate change is threatening to diminish this revenue as corals become bleached and experience higher rates of mortality.

Chen *et al.* (2015) conducted a study to estimate the global economic impact from loss of corals as a result of climate change. They identified three main factors from climate change that impact coral reefs the most: sea surface temperatures, CO_2 concentrations in the water, and sea level rise. In order to assess the impact of these factors on coral reefs, Chen *et al.* used a threshold model in which they found that there are two temperature thresholds that may negatively impact coral reefs. When sea surface temperatures are between 22.37°C and 26.85°C , coral cover may increase; conversely, when sea surface temperatures drop below 22.37°C or rise above 26.85°C, coral cover decreases. Chen *et al.* found that increasing CO_2 concentrations also cause a decrease in coral cover, while sea level fluctuations were found to have no significant effect.

In order to evaluate the value of coral reefs, Chen *et al.* used a meta-analysis that incorporated the percent coral cover, number of visitors to the reefs, GDP per capita, and the tourism expenditure for each visitor. They found that when coral cover decreased, reef value was reduced. The number of visitors correlates negatively with coral reef value because visitors prefer to visit uncrowded coral reefs. The GDP per capita and the tourism expenditure for each visitor were found to have positive effects on coral reef value.

Lastly, Chen *et al.* developed four different mitigation scenarios in response to climate change to evaluate coral reef value. The impact of these different mitigation scenarios on tourism and recreation revenue varies as coral cover varies under these scenarios. The economic loss ranges from $1.88 billion to $12.02 billion by the year 2100. Chen *et al.* noted that this result only represents the coral reef value from tourism and recreation and that there are many other factors that will be impacted by a decline in coral cover; therefore, they create a crude economic loss estimate under these four mitigation scenarios that ranges from $3.72 billion to $23.78 billion.

Overall, CO_2 and sea surface temperatures will affect coral cover, which will reduce the coral reef value. A reduction in coral reef value reduces the recreation and tourism expenditures amongst other factors; therefore, ensuring coral cover remains high will give a higher guarantee that recreation, tourism.

Conclusions

Global warming has been shown to negatively impact corals, as evidenced by the mass coral bleaching event in 1998 and other bleaching events. However, studies have demonstrated that the fate of corals and their ecosystem may not be as ominous as once thought. Many studies have discovered that once corals become bleached they are capable of recovery, while other studies have found that corals may potentially be more resistant to a future bleaching event as they acquire more temperature-tolerant zooxanthellae. Though scientists seem to be optimistic about the survivability of corals, a reduction in corals will still be seen; as a result, coral reef ecosystems will most likely experience a reduction in various species, and the ecosystem services provided to humans will decrease as well. Maintaining corals is not only important for the entire coral reef ecosystem to survive and thrive, but it also benefits humans as many coral reefs provide a source of revenue for different tourism and fishery industries across the globe.

Though these different studies provide work on the current and potential global warming impact on corals around the world, many of them do not take into account the other factors that have the potential to negatively impact corals, including sea level rise and ocean acidification. For future studies, a focus should be placed on combining the affects of sea level rise, ocean acidification, and global warming to gain a better understanding of how corals will respond to such changes and the survivability of corals and their ecosystem.

References Cited

Berkelmans, R., Willis, B.L., 1999. Seasonal and local spatial patterns in the upper thermal limits of corals on the inshore Central Great Barrier Reef. Coral Reefs 18, 219-228.

Berkelmans, R., De'ath, G., Kininmonth, S., Skirving, W.J., 2004. A comparison of the 1998 and 2001 coral bleaching events on the Great Barrier Reef: spatial correlation, patterns, and predictions. Coral Reefs 23, 74-83.

Bozec, Y.-M., Alvarez-Filip, L., Mumby, P.J., 2015. The dynamics of architectural complexity on coral reefs under climate change. Global Change Biology 21, 223-235.

Bozec, Y., Mumbery, P.J., 2015. Synergistic impacts of global warming on the resilience of coral reefs. Philosophical Transactions of the Royal Society B, 370: 20130267.

Chen, P., Chen, C., Chu, L., McCarl, B., 2015. Evaluating the economic damage of climate change on global coral reefs. Global Environmental Change 30, 12-20.

Descombes, P. Wisz, M.S., Leprieur, F., Parravicinin, V., Hiene, S., Olsen, S.M., Swingedow, D., Kulbicki, M., Mouillet, D., Pellissier, L., 2015. Forecasted coral reef decline in marine biodiversity hotspots under climate change. Global Change Biology doi: 10.1111/gcb.12868.
http://www.ncbi.nlm.nih.gov/pubmed/25611594

Erwin, D.H., 1998. The end and the beginning: recoveries from mass extinctions. Tree
13, 344-349.

Erwin, D.H., 1994. The permo-triassic extinction. Nature 367, 231-236.

Graham, N.A., Jennings, S., MacNeil, M.A., Mouillot, D., Wilson, S.K., 2015. Predicting climate-driven regime shifts versus rebound potential in coral reefs. Nature doi:10.1038/nature14140.

Great Barrier Reef Marine Park Authority, 2009. Great Barrier Reef Outlook Report 2009. Great Barrier Reef Marine Park Authority.

Greenstein, B.J., Pandolfi, J.M., 2008. Escaping the heat: range shifts of reef coral taxa in coastal Western Australia. Global Change Biology 14, 513-528.

Hallam, A., Wignall, P.B., 1999. Mass extinctions and sea-level changes. Earth-Science
Reviews 48, 217-250.

Jablonski, D., Raup, D.M., 1995. Selectivity of end-Cretaceous marine bivalve extinction. Science 268, 389-391.

Marzoli, A., Bertrand, H., Knight, K.B., Cirilli, S., Buratti, N., Vérati, C., Nomade, S., Renne, P.R., Youbi, N., Martini, R., Allenbach, K., Neuwerth, R., Rapaille, C., Zaninetti, L., Bellieni, G., 2004. Synchrony of the Central Atlantic magmatic province and the Triassic-Jurassic boundary climatic and biotic crisis. Geology 32, 973-976.

Marshall, P.A., Baird, A.H., 2000. Bleaching of corals on the Great Barrier Reef: differential susceptibilities among taxa. Coral reefs 19, 155-163.

Marshall, P.A., Johnson, J.E., 2007. The Great Barrier Reef and climate change: vulnerability and management implications. Climate change and the Great Barrier Reef. Great Barrier Reef Marine Park Authority and the Australian Greenhouse Office, Australia, 773-801.

McGhee Jr., G.R., 2001. The 'multiple impacts hypothesis' for mass extinction: a comparison of the Late Devonian and the late Eocene. Palaeoegoeography, Palaeoclimatology, Palaeoecology 176, 47-58.

Munday P.L., Jones, G.P., Sheaves, M., Williams, A.J., Goby, G., 2007. Vulnerability of fishes of the Great Barrier Reef to climate change. Climate Change and the Great Barrier Reef; a vulnerability assessment. Great Barrier Reef Marine Park Authority, Townsville, Australia, 357-391.

Raup, D.M., 1986. Biological extinction in earth history. Science 231, 1528-1533.

Sheehan, P.M., 2001. The late Ordovician mass extinction. Annual Review of Earth and

Planetary Sciences 29, 331-364.

Silverstein, R.N., Cunning, R., Baker, A.C., 2015. Change in algal symbiont communities after bleaching, not prior heat exposure, increases heat tolerance of reef corals. Global Change Biology 21, 236-249.

Twitchett, R.J., 1999. Palaeoenvironments and faunal recovery after the end-Permian mass extinction. Palaeogeography, Palaeoclimatology, Palaeoecology 154, 27-37.

Wall, M., Putchim, L., Schmidt, G.M., Jantzen, C., Khokiattiwong, S., Richter, C., 2015. Large-amplitude internal waves benefit corals during thermal stress. Proceedings of the Royal Society B, 282: 20140650. http://dx.doi.org/10.1098/rspb.2014.0650

Winkler, N.S., Pandolfi, J.M., Sampayo, E.M., 2015. *Symbiodinium* identity alters the temperature-dependent settlement behaviour of *Acropora millepora* coral larvae before the onset of symbiosis. Proceeding of the Royal Society B, 282: 20142260.

Effects of Marine Conservation

Weronika Konwent

Marine conservation is a hot topic in today's world threatened by climate change. There are arguments from conservationists and political figures both for and against it, a paradox unusual in the scene of conservation. The government advocating for no-take marine zones in national waters? Conservationists protesting the creation of huge areas of protected ocean? Obviously there are different motivations for marine conservation, which lead to controversy in designation, implementation, and effectiveness of marine protected areas (Singleton and Roberts 2014). Researchers have been studying the effects of this and how better to approach these issue in the future, in a world that will presumably need intervention more than ever.

In order to mitigate and rectify inefficient and nonexistent marine conservation practice, researchers are approaching the issue from many angles. Not only must abiotic factors of spatial distribution and water chemistry be observed, but also biotic factors of biodiversity, adaptation, and acclimation. Additionally, this must be set within a framework of human social and economic costs. This entire system must then be extrapolated into the future, taking into account all of the potential hazards of climate change. Ideally, such models must also be effective for widespread geographic areas, as well as in locations with poor data. This is understandably difficult, and conceivably close to impossible to achieve. Protecting ocean biodiversity in space and time without stepping on international toes or disturbing patterns of local and national economy will require concessions that neither party may be willing to make. Meanwhile, sea temperatures and levels will continue to rise, pushing species away from natural habitats and shifting the balance in marine ecosystems. If these habitats are to survive, something will have to be done, and at least in terms of this controversy some action is better than no action at all.

Are No-Take Marine Reserves Really Effective?

No-take marine reserves (NTMRs) are established to promote marine biodiversity and to reign in exorbitant fishery behavior through the prohibition of all fishing and resource extraction. In the past several decades, the quantity of NTMRs has risen greatly across the world. While this seems like an obviously positive trend, many regard the existence of NTMRs as controversial. While most NTMR results appear positive, the opposition claims that faulty study design and a lack of objective empirical data may cause inaccurate portrayals of NTMR effects. Miller and Russ (2014) test these claims to better understand the trend NTMRs are taking.

The main trouble in the investigation of NTMRs is the analysis of the differences in habitat between NTMRs and fished sites. In order for results to be accurate, the sites under observation must be as similar as possible. There are five main methodologies that researchers use to study NTMR effect. In method (a) entire habitats are tested for similarity, but the specific sampling sites are not described. This is problematic due to diversity even within cohesive habitats, and is often associated with biased results. Method (b) is the most common method by which to investigate NTMR effectiveness, and was used by almost half of the studies reviewed in this article. This method has the opposite problem of method (a), as it selects similar sampling sites that are not necessarily within similar habitats. This provides more accurate but not unequivocal results. Method (c) statistically tests whether a difference exists between the two habitats, and if they are not significantly different it is assumed that there is enough similarity to warrant research. This method is the most effective, but only contingent on the inclusion of very specific habitat factors. Method (d) is a more detailed version of method (c) in which the habitat factors are specifically tailored for habitat variability, and can therefore be removed from statistical testing. The descriptiveness of this method is beneficial, but only accurate when habitats are measured over time or in conjunction with baseline data. Finally, method (e) studies habitats as a result of NTMRs, but does not take into account the natural variation that would occur over time. This overly simplistic view does not provide clear enough data from which to draw conclusions of NTMR effectiveness.

These five methods of research must then be applied to a sampling design. There are three designs considered in this article. Single-point-in-time studies compare data from a single point in time. They are the most limited measure of effectiveness due to their limited viewpoint. Temporal monitoring studies monitor the changes found in NTMRs since their creation. They are better than single-point-in-time studies, but still lack baseline data that would make their conclusions concrete. Before-after-control-impact studies monitor changes found in habitats pre- and post-NTMR implementation. These studies provide the most robust data as they measure both habitat differences between sites and habitat change over time.

Ultimately, NTMR research should be considered with caution. Even the most carefully designed studies may not accurately portray the effects of these marine reserves, and poorly designed studies may be completely misleading. They might either provide too hopeful an estimate, which may result in ill-placed investment in NTMRs, or they might provide too pessimistic an estimate, which may reduce faith in the effectiveness of NTMRs. Either way the research will be incomplete, which not only makes interpretation more difficult, but is also plain bad science.

Addressing Climate Change in Australian Marine Ecosystems

Australia's diverse marine environment is under threat from varied effects of climate change such as marine heat waves, ocean acidification, floods, and tropical cyclones. Various organisms spanning many habitats are affected, including fish, seabirds, marine turtles, coral, and marine invertebrates, many of which are keystone species that influence the structure of a particular community. Johnson and Holbrook (2014) recognize the importance of studying and striving to understand the impact of ecological changes on the habitat and its inhabitants in order to preserve them as effectively as possible.

The study of climate change effect on marine habitats is shifting towards a study of thresholds at which a small change in conditions can have a massive effect on the ecosystem. Effects of such thresholds are often very sudden and extremely difficult to bounce back from, so emphasis is being placed on avoiding thresholds, and instead focusing on resilience techniques that allow organisms to forestall major environmental shifts. Management of local pressures is an effective strategy used to improve ecosystem resilience, but the dynamic and delicate nature of marine habitats precludes the possibility of using this as an absolute solution. While there exist many preservation methods, the most politically popular and easy to integrate is the marine protected area. MPAs have long shown a history of effectiveness, yet as a response to climate change they may leave something to be desired. Their lack of direct address of climate change factors detracts from their utility. Limitations on human pressures exist, but are not being enacted effectively enough to impact ecosystem resilience. The rapid rate of climate change also limits the feasibility of long-term projects, which may mean that researchers will simply have to trust in the adaptive properties of genetic diversity. As climate change places additional stress upon already weakened marine habitats, MPAs will be effective only as supplemental protection against everyday pressures. Instead, a more comprehensive program that encompasses ecological resilience, ecosystem protection, and anthropogenic limits will need to be adopted.

A Strategy for Response to Climate Change in Marine Conservation

An effect of global warming is an increase in sea-surface temperatures (SST), which impacts the distribution and range of corals. As temperatures increase, coral distribution will shift poleward. This is problematic because current marine protected areas do not take into account the distribution effects of climate change. Continual shifting of MPAs as conditions worsen is more than likely to meet political and logistical roadblocks. Makino *et al.* (2014) established an integrative system by which to determine priority selection of habitats for MPAs. This research aims to create a process through which climate change can be factored into subsequent MPA planning, and will cater to coral distribution trends not only now but in the future as well. A program called Marxan is used to design MPAs that will be most cost efficient, and therefore most likely to be implemented, and most connective in terms of time and space. The benefit of such an approach is its versatility—Marxan can be used to create sites in any place, at any geographic scale, and at any time. This is powerful because of the usability of such a tool by governments and organizations worldwide. With it, researchers hope that the designation of effective MPAs, even when faced with lack of substantial data, can prove up to the job of tackling the dynamic, rapid, and destructive effects of climate change on the marine environment.

Conservation Strategies for a Changing Climate

Conservation of marine species, especially as a response to climate change, requires a reliable conception of current and future spatial distribution of species to allow for the protection of biodiversity and the establishment of conservation at the most appropriate sites. Gormley *et al.* use Species Distribution Modeling (SDM) to predict how Priority Marine Habitats (PMH) in the NE Atlantic might shift and change according to climate change induced changes.

In this study, predicted changes were used to draw inferences on many different topics, including the spread of PMHs both now and in the predicted climate change-affected world 100 years from now, the overlap between PMHs and Marine Protected Areas (MPAs), whether MPAs are equipped to accommodate PMHs, and how this will affect conservation and management planning in the future. With current trends aiming towards expansive established international protection, steps are being taken to predict anthropogenic interaction with MPAs and how a balance can be maintained. In this study, several independent factors were chosen to portray a suitable environment based on the ecological requirements of a typical PMH. A SDM program was then used to designate potential PMHs. These areas were then compared with existing MPAs, and crossed with commercial zones. These results are meant to provide a map of potential PMH hotspots and a framework for future MPAs. Inaccurately placed MPAs have the potential to aggravate fragmentation, which, combined with habitat loss from climate change, may lead to serious issues with population response and adaptability. As suitable species-specific habitats move and change, a confident model of species distribution that supports effective creation of protected areas will help to support the natural adaptive processes that regional species will have to undertake to keep up with the times.

Intersection of Biodiversity and Socioeconomic Interests

Due to lack of specific species data, it is often difficult to predict where marine conservation will be most effective in maintaining biodiversity and ecosystem functionality. Olds *et al.* (2014) test whether surrogates that fulfill the criteria of being keystone, umbrella, and flagship species can accurately predict which areas are optimal for conservation. They also tested whether seascape connectivity has an effect on fish abundance. It was concluded that the integration of these two conditions in marine spatial planning can positively impact the maintenance of fish communities and the functioning of ecosystems, and that these improvements can be beneficial to people in terms of sustenance and income.

Bumphead parrotfish were chosen as surrogates to map multi-species conservation. Because they fit the criterion of threatened, crucial for the ecosystem, important to socioeconomic conditions, and share a pattern of movement with many different species, they can be used to plot marine conservation areas with a correlated effect for many species. Seascape connectivity between seagrass, mangrove, and coral reef environments was also used to predict the ability of reserves to enhance effective conservation. This is due to the variety of habitats the bumphead parrotfish and consequently many other local species utilize throughout their lifetimes. Divers observed the distribution of species throughout specific sites, and then compared the results in terms of species abundance and distribution. Results showed that connectivity improved species abundance, and that reserves that benefitted the bumphead parrotfish also strongly benefitted other local species. These results support the use of seascape connectivity and surrogate species in spatial planning to promote the functioning of ecosystems and abundance of other species. This is important not only for biodiversity, but also for the socioeconomic effects that more abundant flagship species can provide. A win-win for people and the environment is a strong point in the case for thoughtful marine conservation.

Two Types of Science, One Study of Ocean Acidification

Ocean acidification is predicted to increase as global warming accelerates, affecting marine habitats and especially coastal areas experiencing episodic upwelling, such as the California Current Large Marine Ecosystem (CCLME). Hofmann *et al.* (2014) are studying this particular habitat due to its wide variety of conditions and its particular susceptibility to rapid environmental change, To do this, they are using data collected by the Ocean Margin Ecosystems Group for Acidification Studies (OMEGAS) to pair oceanographic and biological data to create a more thorough understanding of genetic variability within key species populations, and how this can affect adaptation to the conditions caused by climate change. Using the biological data to measure responses of sea creatures to oceanographic factors that are affected by climate change, Hofman *et al.* (2014) hope to plot the future survival of CCLME species.

Measurements of pH, salinity, and temperature, along with others, are taken as oceanographic data to describe the abiotic conditions of the habitat. The benefit of the CCLME is that it spans a large distribution of these conditions, which allows researchers to approximate change in time as change in space. Concurrently, biological data such as growth rate, calcification, and physiology, and genetic data are used to measure organisms' responses to these varying conditions.

The specific organisms tested are key species of calcifying benthic marine invertebrates. By researching the responses of these organisms, researchers hope to gain a better understanding of the genetic variability within populations, and how this might help or hinder species as global warming causes ocean acidification. This particular method of investigation has several benefits. It tests the possible location of refuge sites that are likely to serve as safety reserves against ocean acidification, is economical for scientists to employ, and allows the study of sprawling ecosystems. In this particular study, it was found that benthic marine invertebrates have a high resilience to such changing conditions, and that local adaptation can occur through space and presumably time, although it is not completely certain whether organism response can be attributed solely to the effects of ocean acidification. Nonetheless, OMEGAS has begun to use this method to test for acclimation and local adaptation in such sites globally, and to predict the effects of ocean acidification on a diverse variety of species. Such knowledge can then be used in the creation of marine reserves, or to communicate with local policy makers and scientists, which can have far-reaching and positive effects.

Marine Conservation: How to Reconcile Fish and Food

The growing pressure to establish marine protected areas (MPAs) in response to climate change, anthropogenic involvement, and the existence of endangered species is matched only by the economic restrictions such programs are bound by. Mazor *et al.* (2014) tackle this problem with a system that focuses specifically on opportunity cost, as opposed to area, to most effectively plan for efficient conservation. They found that the identification of priority conservation areas is best achieved by using cost data rather than biodiversity data, highlighting the importance of incorporating such information in conservation efforts, especially where multiple countries are involved.

For the area of interest, Mazor *et al.* (2014) chose to study the Mediterranean Sea, an area of marine importance to many countries, for many reasons. The effect of marine conservation on economy and culture within countries that possess different socioeconomic statuses, styles of government, and international relations varies widely, but must be considered in areas such as the Mediterranean that are shared internationally. In this case, within the question of opportunity cost the researchers examined commercial fishing, noncommercial fishing, and aquaculture. The target was set to protect 10% of each of the 77 threatened marine species within the area with minimum opportunity cost. The results of the experiment showed that when opportunity cost was incorporated into conservation planning, percentage of annual income necessary to meet objectives was reduced, and that priority areas, cost, and area requirements differed from country to country. This is particularly important as recent efforts to establish MPAs are led by biologists and conservationists who are not necessarily well versed in the realm of economics. Also, the discovery of the commercial fishing sector as a lead player in determining the opportunity cost of marine reserves especially dictates the placement of priority areas for conservation. While these conclusions are telling, if they are to be implemented issues of illegal fishing, political stability, and international relations will also have to be addressed, which may require not only scientists, but diplomats as well.

VLMPAs: Gain or Loss?

The creation of very large marine protected areas (VLMPAs) is trending, at least in political circles. Conservationists are skeptical of the effectiveness of such sizeable marine protected areas, and argue both against the criteria used to establish them and their ability to be successfully policed. Singleton and Roberts (2014) conclude that no matter the effectiveness of VLMPAs, the publicity they bring to environmental efforts is valuable, and may lead to sustainable conservation practices in the future.

VLMPAs, classified as reserves over 100,000 km^2, make up more than half of the area protected in MPAs worldwide as of 2012. With a goal of protecting 10% of the world's oceans by 2020, VLMPAs are playing an important role in reaching the coverage necessary. Common arguments against VLMPAs include the lack the diversity of smaller areas, and being unable to guard against dangers of localized disasters. The vast size of these reserves refutes these arguments, both by spanning multiple habitat types and by outsizing even the most immense disasters. A downside of VLMPAs is the reduced benefit of edge to area in spillover effects. This limits the commercial use of protected areas for fisherman, which may lead to lack of compliance. Conservationists also condemn catering to resource users, with a particular emphasis on those VLMPAs skirting the most valuable fishing, oil, and gas areas. Establishment of current VLMPAs far from major human populations leads to a failure in confronting anthropogenic issues of over-fishing, pollution, and tourism. By trying to create marine reserves with minimum resistance and cost, some say VLMPAs sacrifice quality for quantity. Ultimately the creation of VLMPAs does not detract from the creation of smaller, coastal MPAs, nor does it slow the process of conservation. Instead it creates heated dialogue of effectiveness of marine conservation on a global scale, and if that isn't a win for our oceans VLMPAs won't make a difference, colossal size or no.

Conclusions

Many different groups and organizations are tackling the issue of marine conservation from different perspectives. While some focus on the traditional question of biodiversity, others create models, and still others examine social and economic costs. As marine conservation becomes more urgent, perspectives will presumably begin to align, and what is a controversy today will become the obvious next step of tomorrow.

References Cited

Gormley, Kate S.G., Hull, Angela D., Porter, Joanne S., *et al.* 2014. Adaptive management, international co-operation and planning for marine conservation hotspots in a changing climate. Marine Policy 53, 54-66.

Hofmann, G. E., Evans, T. G., Kelly, M. W., *et al.* 2014. Exploring local adaptation and the ocean acidification seascape – studies in the California Current Large Marine Ecosystem. Biogeosciences 11, 1053-1064.

Johnson J.E., Holbrook N.J., 2014. Adaptation of Australia's Marine Ecosystems to Climate Change: Using Science to Inform Conservation Management. International Journal of Ecology 2014. doi:10.1155/2014/140354

Makino, A., Yamano, H., Beger, M., *et al.* 2014. Spatio-temporal marine conservation planning to support high-latitude coral range expansion under climate change. Diversity and Distributions 20, 859–871.

Mazor, T., Giakoumi, S., Kark, S., *et al.* 2014. Large-scale conservation planning in a multinational marine environment: cost matters. Ecological Applications 24, 1115-1130.

Miller, K. I., Russ, G. R., 2014. Studies of no-take marine reserves: Methods for differentiating reserve and habitat effects. Ocean and Coastal Management 96, 51-60.

Olds, Andrew D., Connolly, Rob M., Pitt, Kylie A., *et al.* 2014. Incorporating Surrogate Species and Seascape Connectivity to Improve Marine Conservation Outcomes. Conservation Biology 28, 982-991.

Singleton, R.L., Roberts C.M., 2014. The contribution of very large marine protected areas to marine conservation: Giant leaps of smoke and mirrors? Marine Pollution Bulletin 87, 7-10.

Effects of Climate Change on Agriculture

Adin Bonapart

The agricultural sector is a large contributor of greenhouse gasses (GHG) and is, conversely, highly vulnerable to changes in climate. Rising global temperatures are shown to result in a net loss of crop yields around the world. Hotter temperatures will especially affect equatorial regions, where changes in local climate conditions may exceed the temperature threshold of certain crops, which already receive a maximum amount of solar gain (Liu *et al.* 2005). The projected rising of potential evapotranspiration (PET) levels, caused by hotter temperatures, is shown to result in significant increases in water loss in agroecosystems and lead to increased drought conditions around the globe. Thus, general trends of climate change and agriculture show a shift of available growing regions away from the equator and toward the poles (Scheff and Frierson 2014).

A globally-expanding human population, a hotter climate, and inefficient farming practices translates to greater food insecurity around the world (Gregory, Ingram, and Brklacich 2005). Feeding more people and in hotter conditions will be very expensive, requiring more farmland and ecologically unsound infrastructure such as dams. Furthermore, there is a studied correlation between economic status and dietary efficiency that results from consumer demand for more energy intensive food (meat/dairy) consumption at higher income levels (Tilman *et al.* 2011). Although agricultural lands cover almost half of earth's habitable land, farms support very little biodiversity and ecosystem services. Modern inefficient agricultural practices lower topsoil, reduce biodiversity, require enormous water and energy inputs, and increase atmospheric CO_2 through the conversion of native ecosystems to farmland, particularly in the tropics. Inefficient farming practices and consumer demand are two important factors that contribute heavily to agriculture's carbon footprint and make it particularly susceptible to changes in climate.

Food security and a reduction in agriculture-related emissions can be achieved via sustainable intensification of agriculture, via supply and demand side reductions. Supply side reductions include methods such as crop rotation and companion planting, reduced tilling, improved irrigation and fertilization efficiency, elimination of fossil fuels or synthetic pesticides and fertilizers, and the incorporation of crop wild relatives (CWR). Demand side reductions include food waste reduction from farm to fork, and the incorporation of healthier and climate-friendly diets from peoples in wealthier countries. Models show that reducing food agricultural waste by half greatly reduces the area of cropland and GHG emissions associated with agricultural systems (Bajželj, et al. 2014). Adoption of healthier diets, as a reduction in the consumption of meat, dairy, and other energy-rich food commodities by the industrialized nations, are also considered an essential facet of climate mitigation via the agricultural sector. There is a tiny amount of food delivered compared with the overall productivity associated with livestock systems, with energy losses at every step of production; making animal agriculture a very inefficient form of land usage. Livestock, especially large grazing animals like cattle, currently accounts for about one-third of all the GHG emissions from agriculture, and over 14% of all human-caused GHG emissions worldwide (FAO, 2015, Bajželj et al. 2014).

Fortunately, there is a scientific consensus that food security and agriculture emissions reductions can be achieved if farmers and policy makers adopt tenets of sustainable intensification. Important research is being conducted on adapting agriculture to climate change, including the adoption efficient farming techniques, healthier diets, and heat-tolerant CWR.

PET to Increase with Greenhouse Warming

A 2014 study by Scheff and Frierson found that potential evapotranspiration (PET) from soil-vegetation systems, the rate at which surface water evaporates if available in a given climate, is projected to rise with greenhouse warming, leading to increased drought conditions around the globe. Projected annual-mean PET fields for 2089–2099 were analyzed and found to be almost always positive, with typical increases of 10%–45%, compared with 1981–1999 levels. To obtain these results, Scheff and Frierson calculated changes in annual-mean PET using the Penman-Monteith Equation, a fundamental physical quantification containing many different climatological variables, and compared the results across 13 different global climate models.

Biologically, evapotranspiration in plants is a necessary "cost" of opening the stomata to permit the diffusion of CO_2 gas for photosynthesis. This important phenomenon also cools plants and allows the transfer of nutrients and water from the roots to the shoots. Quantifying PET has important implications for the agricultural sector because higher PET climates are more arid, drought-prone, and evaporative climates.

Scheff and Frierson maintain that the main reason for PET increases is from warming itself (single digits of °C in most places), rather than other greenhouse factors (such as increases in surface net radiation). Rising temperatures was found to cause the PET increase by widening the vapor pressure deficit variable in the Penman-Monteith Equation. This was shown mathematically, as well as graphically by isolating only ambient air temperature for 2081–2099 and leaving all other climatological variables at 1981–1999 levels. Model results differ because of variations between inputs, and the exact amount of warming is a major source of uncertainty in projected PET levels.

Global Crop Losses from Climate Change, GGC Models Predict

Rosenzweig *et al.* (2014) perform an intercomparison of seven global gridded crop models (GGCMs) and analyze their combined effectiveness in predicting the outcomes of climate change on the world's food systems. Differences in the structure, purpose, and process between individual models are major sources of uncertainty in multi-model climate change assessments. Regardless of their high degrees of uncertainty however, global crop projections indicate substantial crop-yield losses worldwide, particularly in the tropics.

Although climate change will affect agriculture globally, the strongest effects are going to be felt in the lower latitudes where developing countries are concentrated. Even moderate (1–2°C) rises in temperatures will cause significant reductions in agricultural yields in tropical regions, which receive maximal solar energy. Elevated atmospheric temperatures will result in increased evapotranspiration and water demand for crops in these regions, where many important cereal crops are already at or near their upper temperature thresholds.

Nitrogen concentration is another key determinant in the effects of changing climate on crop yields because it limits the photosynthetic enhancement effects from elevated CO_2 levels. Simulations that account for nitrogen stress levels show dramatic effects, resulting in over 50% loss of yield in some tropical regions for maize, wheat, rice, and soy, the world's four top agricultural products.

CWR and Global Food Security

Domestication by humans has reduced the genetic diversity within certain crops over time, making agriculture more susceptible to changes in climate. Some of the relevant effects of global climate change include shifts in temperature, rain variability, and plant pathogen range, all of which impact crops in various ways. Models predict that such climate-driven effects account for yield losses of 6 to 10% per 1°C of warming (Guarino and Lobell 2011). Furthermore, a global human population predicted to reach over 9.3 billion by 2050, plus degraded soils, water, land, and other resources, is creating further instability for food systems around the world.

"Adapting Agriculture to Climate Change" is a ten-year project supported by the Government of Norway and managed by several global seed banks, which is addressing global food security by creating a collection of major crops and their relatives with the characteristics required for adapting to climate change. The 2014 paper by Dempewolf *et al.* details the goals, methods, and results of this on-going project.

The researchers consider the crop wild relatives (CWRs) as valuable pools of genetic diversity, in which to find the tools for adapting agriculture to climate change. Historically, the crossing of crops with CWRs has been used for agricultural purposes such as increased yields and adapting plants to drought, cold and increased saline conditions. This project will systematically collect, conserve, and evaluate CWRs of 29 focal crops under predicted climate change conditions, and identify useful traits and make this material available for agricultural use.

According to the study, there is very limited CWR gene pool data, and no global assessment of the state of conservation of CWRs. The first completed step of the project was the collation of CWR taxa for over 173 crop gene pools into a checklist and to make this data available online. The researchers estimate 10,000 species of CWR of "high potential value" to food security. The project is time sensitive however, because most of the relevant CWRs are threated in their wild ecosystems by human development, and climate change is likely to further compound these pressures.

Dams and Agriculture in Idaho

Water storage and distribution infrastructure (dams) allow large areas of land that wouldn't otherwise have access to water (i.e. away from riparian areas) to be farmed and settled. Furthermore, dams give farmers security against variations in climatic conditions and water supply (i.e. droughts), which, allows farmers to grow higher-valued, more water intensive crops. Hansen *et al.* (2014) find that the presence of dams has a "small, positive, but non-significant effect" on farmland values. For these reasons, the construction of dams tends to lead to improved crop yields and planted acreage.

Hansen *et al.* combine spatiotemporal data to examine the impact of major dams on irrigated agriculture and the natural environment in Idaho. In many regards, this study is successful in illustrating the relationships and interdependency between modern agriculture and major water storage projects. Irrigated agriculture is responsible for about 85% of all water withdrawal in Idaho, and about half of all major dams in the state list irrigation as one of the essential purposes. Overall, the authors argue that the development of major dams is beneficial for the agricultural sector, although they maintain that dams are generally detrimental to the natural environment.

Agricultural irrigation practices and large dams are connected with negative ecological impacts as a result of diverting and depleting water from streams and rivers. Such artificial changes to the watershed affect fish and other wildlife, degrade water quality by changing the salinity of the water, and disrupt natural flows that can have devastating consequences native fish populations. But, Hansen *et al.* 's assessment on dams and the natural environment is lacking in these quantitative data. The study gives measurements on annual and seasonal stream flows along a major river in Idaho and touches on the intersection of ecological problems and agricultural land use, but it does not attempt to quantify any ecological data. While, in the title and in the abstract, the authors claim to evaluate the comprehensive long-term impacts of dams on the natural environment, they do not attempt to incorporate any *ecological* costs of dams into their models, nor do they investigate the connection between the integrity of aquatic ecosystems and the sustainability of these water supplies.

That the authors didn't explore the ecological affects of dams and agriculture is emblematic of the current conceptual rift between human behavior and ecological relationships that is present in agricultural and natural resource systems today. The current water "regime" in Idaho is built upon antiquated 1800s conceptions of property, including the priority in appropriation (first in time first in right) clause, which is a highly human-centric and problematic land ethic in the context of a globalized world characterized by ecological crises and resource mismanagement. Hansen *et al.* are vastly limiting the usefulness of their study when they consider agriculture and the natural environment as two separate "sectors," rather than one system of interconnected parts. If the authors had considered the long-term ecological impacts of dam construction and mass irrigation projects, their conclusions might have been significantly different.

Reduced Tillage Decreases Erosion and Runoff, Uruguay Tomato Study Finds

A 2014 study by Alliaume *et al.* found that implementing conservation agriculture (CA) techniques such as tillage reduction or the inclusion of mulch and chicken manure could decrease soil erosion and agricultural runoff by more than 50%, compared with conventional methods. The two-year study monitored the changes in yield, erosion, and water dynamics of four different treatments of a tomato-oat rotational crop in a temperate climate in Uruguay. The different treatments, which were all in permanent raised beds included: conventional tillage systems with only artificial fertilizer as a control (*CT*), a mixture of chicken manure and rice husk (*CChM*), a combination of both chicken manure and green manure consisting of black oats (*CGM*), and a reduced tillage treatment (*RT*), which also incorporated chicken manure and black oat mulch.

According to the study, reduced tilling and mulch increases the water threshold at which runoff starts by 49%, resulting in greater water capture and storage compared to conventional methods. Further, the presence of a cover crop is found to reduce soil loss by 98%. Interestingly, fruit productivity was greatest under the *CChM* treatment (conventional tilling with chicken manure) for both years, with more than 50% greater yield than under *RT*. According to the researchers, the low yields under the *RT* treatments are ascribed to a combination of poor crop establishment under the black oat cover crop and nitrogen immobilization, which was also observed in the CGM treatment (conventional tilling and black oat as green manure).

The nitrogen levels for the two beds containing the oat were found to be at or below the critical value limit for nitrogen. The researchers speculate that the large amounts of carbon-rich cover crop in the soil led to nitrogen immobilization, affecting the supply of soil nitrogen to the crop.

Organic Farming Safer for Farmers' Health, Portuguese Study Finds

A 2014 study by Costa *et al.* found that farmers who do not use pesticides demonstrate lower levels of DNA damage and immunological alteration than their pesticide-using counterparts in the conventional agricultural sector. Biomarkers of pesticide exposure, effects, and susceptibility were evaluated using urine samples and genotype analysis of three different groups under study: pesticide farmers, organic farmers, and a non-exposed control. The study concluded that pesticide use leads to increased DNA damage in farmers, although differences in exposure conditions may influence results. Overall, organic farmers demonstrated a level of genetic damage similar to the unexposed controls.

Variables that may influence the effects of pesticide exposure on the human body include host factors such as gender, age, lifestyle, working environment, as well as application frequency and in/adequate usage. Concerning pesticide *exposure*, Costa *et al.* found that the concentration of organophosphate and carbamate compounds in urine was the only marker that yielded significant differences between the three groups under study. Both pesticide exposure *and* effects are elevated during the spring-summer period, when maximum pesticide usage in agriculture occurs. Improper handling and usage of pesticides also greatly influences both of these levels. Likewise, workplace conditions have shown to be an extremely significant factor on exposure risk, and greenhouses in particular are known to increase the risk of pesticide exposure. Interestingly, while the Costa *et al.* study confirmed that greenhouses are associated with higher pesticide exposure, greenhouse farmworkers were found to present *lower* levels of DNA damage. Costa *et al.* found no significant differences of pesticide concentrations between different age groups, or between smokers and non-smokers. These findings contrast with Bolognesi *et al.* 2002, and Costa *et al.* 2006, which did, in fact, find increased genetic damages to greenhouse farmworkers.

Biodynamic Wine is More Sustainable, Spanish Study Finds

A 2014 study by Villanueva-Rey *et al.* found that environmental impacts from viticulture are substantially reduced with biodynamic practices, as opposed to conventional methods. The environmental impact reductions obtained from biodynamic grape-production systems are attributed to an 80% decrease in diesel fuel, pesticides, fertilizers, and other external inputs. Diesel use is the main source of environmental impacts for viticulture, and is up to 4 times lower in biodynamic systems resulting from differences between mechanized farming techniques in conventional vineyards and the implementation of artisanal methods in biodynamic wine-growing.

The Villanueva-Rey study deploys a life cycle impact assessment (LCIA) of three different viticulture techniques in Northern Spain: biodynamic cultivation, conventional vineyards, and an intermediate biodynamic-conventional wine-growing plantation. The environmental impacts assessed in this study were: abiotic depletion potential, acidification potential, eutrophication potential, global warming potential, ozone layer depletion potential, photochemical oxidant formation potential, eco-toxicity, and land competition. Compared with conventional viticulture practices, biodynamic methods show substantially lower environmental profile for *all* of the impact assessment categories, other than land competition.

Villanueva-Rey *et al.* assessed the global warming potential (GWP) by measuring the annual green house gas (GHG) emissions between the three test sites. The biodynamic test site GWP measured 71.11 g CO_2 eq./bottle wine and the conventional site was measured at 238.42 g CO_2 eq./bottle for the year 2011, a difference of almost 400%. This trend is similar across literature values for other wineries as well.

Natural and synthetic pesticides also make significant contributions to the environmental profile of viticulture systems. Villanueva-Rey *et al.* found that synthetic pesticides such as folpet or terbuthylazine account for 99% of the total environmental burdens for the eco-toxicity category for conventional vineyards. In contrast, pesticide usage for the biodynamic and the biodynamic-intermediate wine-growing sites represented 9% and 20% of the eco-toxicity impact categories respectively. These two sites did not deploy synthetic pesticides.

Although biodynamic viticulture requires more land, human labor, and has lower annual grape yields than conventional vineyards, biodynamic viticulture could be considered as a future climate mitigation strategy for the wine sector because it requires fewer external inputs, lower production costs, and minimizes environmental impacts.

Healthier Diets Needed to Avert Climate Change

Using global datasets, the Bajželj *et al.* (2014) study models different agriculture-based climate mitigation scenarios that minimize the expansion of cropland while insuring global food security. The business-as-usual (BAU) projections for 2050 result in a scenario in which global agriculture alone produces ~21 gigatons of CO_2 every year, almost the full 2°C global target emissions allowance in 2050. The study quantifies the loss of Net Primary Production potential along the agricultural biomass flow, and identifies areas of significant food waste and inefficient farming practices for improvement. The researchers then examine the effects of different "demand-side" or "supply-side" agricultural efficiency measures and solutions.

Supply-side management changes include what the study identifies as yield-gap closures, which behave as if agricultural land is intensively farmed in ways that improve yields as well as reducing environmental impacts (i.e. improved irrigation efficiency and eliminating over-fertilization). According to the models, if sustainable intensification of agriculture were implemented worldwide, about 20% additional crop and pasturelands would still be required to feed the global human population in 2050 and GHG emissions would increase by 42% (from 2009 levels). As such, supply-side measures alone are not sufficient, nor will they provide sufficient agricultural GHG reductions and avoid dangerous levels of climate change.

Primary demand-side measures include food waste reductions and alterations in diet of peoples from richer countries. Wasted food is costly because it has accumulated previous transformation stages that required numerous inputs of resources and energy. Improving crop storage and implementing waste reductions programs are one conceivable way of reducing emissions from agriculture worldwide. Models show that, by reducing food agricultural waste by half, the area of cropland required is reduced by about 14% and GHG emissions are reduced by 22–28% (compared with their respective baseline scenarios for 2050).

The Bajželj et al. study introduces the concept of dietary adaptation, or "Healthy Diets," as a demand-side measure for mitigating emissions and re-allocating resources. Healthy Diets, as a reduction in the consumption of livestock products (meat and dairy) and other energy-rich food commodities by the industrialized nations, is considered an essential facet of climate mitigation via the agricultural sector. Scenarios including Healthy Diets significantly reduced the area for cropping/pasture and GHG emissions by ~45%, with almost all of the large GHG emissions associated with livestock reductions. There is a tiny amount of food delivered compared with the overall productivity associated with livestock systems, with energy losses at every step of production; making animal agriculture a very inefficient form of land usage. Additionally, allowing retired crop/pasture lands return to more natural states of vegetation allows for CO_2 sequestration and encourages better health by removing the overabundance of energy rich foods from the diets of people from the richer countries. Demand-side reductions in this area can be best attained with economic incentives (i.e. a carbon tax), and that the livestock sector should enter a comprehensive climate mitigation policy.

A combination of supply and demand-side solutions are required to avoid dangerous climate change and ensuring global food security.

Studies like this thus challenge the foundations of the modern industrialized food system, demanding that health requirements dictate
agricultural priorities, and not the other way around.

Conclusions

The agricultural sector is expected to undergo major changes if
it's going to adapt to hotter, drier temperatures, a changing energy
paradigm, and an expected 9.6 billion people by 2050. Sustainable
intensification of agriculture will usher the development and implementation of new and old efficient farming practices. Farms must
become more biologically diverse, safer, smaller, more productive,
and energy efficient. Politicians interested in feeding global populations must account for the social costs of carbon and incentivize the
transition to a sustainable agricultural system. Global food security
and climate change is not just an agriculture issue but will involve
governments, researchers, consumers, and producers.

References Cited

Alliaume, F., Rossing, W., Tittonell, P., Jorge, G., Doglioitti, S.,
2014. Reduced tillage and cover crops improve water capture
and reduce erosion of fine textured soils in raised bed tomato systems. Agriculture, Ecosystems and Environment 183, 127-137.

Bajželj, B., Richards, K., Allwood, J., Smith, P., et al. , 2014. Importance of Food-Demand Management for Climate Mitigation.
Nature Climate Change 4, 924-29.

Bolognesi, C., Perrone, E., and Landini, E., 2002. Micronucleus
monitoring of a floriculturist population from western Liguria,
Italy. Mutagenesis 17, 391–397. Costa, C., García-Leston, J.,
Costa S., Coelho, P., et al. , 2014. Is organic farming safer to
farmers' health? A comparison between organic and traditional
farming. Toxicology Letters 230(2), 166-176.Costa, C., Teixei-
ra, J.P., Silva, S., Roma-Torres, J., et al. , 2006. Cytogenetic and
molecular biomonitoring of a Portuguese population exposed to
pesticides. Mutagenesis 21, 343–350.

Dempewolf, H., Eastwood, R., Guarino, L., Khoury, C., et al. ,

2014. Adapting Agriculture to Climate Change: A Global Initiative to Collect, Conserve, and Use Crop Wild Relatives. Agroecology and Sustainable Food Systems 38, 4, 369-377.

FAO. *Major Cuts of Greenhouse Gas Emissions from Livestock within Reach.* Web. 12 Apr. 2015.

Guarino, L., Lobell, D.B. 2011. A Walk on the Wild Side. Nature Climate Change 1, 374-375.

Hansen, Z., Lowe, S., Xu, W. 2014. Long-term impacts of major water storage facilities on agriculture and the natural environment: Evidence from Idaho (U.S.). Elsevier Science Direct, Ecological Economics 100, 106-118.

Gregory, P., Ingram, J., Brklacich, M. 2005. Climate Change and Food Security. Philisophical Transactions of The Royal Society B 360, 2139-2148.

Liu, Z., Vavrus, S., He, F., Wen, N., *et al.*, 2005. Rethinking Tropical Ocean Response to Global Warming: The Enhanced Equatorial Warming. American Meteorological Society 18(22), 4684-4700.

Rosenzweig, C., Elliott, J., Deryng, D., Ruane, A., *et al.* (2014). Assessing agricultural risks of climate change in the 21st century in a global gridded crop model intercomparison. Proceedings of the National Academy of Sciences of the United States of America, 111(9), 3268-3273.

Scheff, J., and Frierson, D. 2014. Scaling Potential Evapotranspiration with Greenhouse Warming. Journal of Climate 27, 1539-1558.

Tilman, D., Balzer, C., Hill, J., Befort, B., 2011. Global food demand and the sustainable intensification of agriculture. Proceedings of the National Academy of Sciences 108(50), 20260-20264.

Villanueva-Rey, P., Vazquez-Rowe, I., Moreira, M., and Feijoo, G., 2014. Comparative life cycle assessment in the wine sector: biodynamic vs. conventional viticulture activities in NW Spain. Journal of Cleaner Production 65, 330-341.

Carbon Storage Capacity of Altered Forest Ecosystems

Stephen Johnson

Across the globe, forests are being destroyed at increasing rates for timber, agriculture, and myriad other human purposes. The removal of forests releases carbon dioxide and reduces the ability of the land to store carbon, which make deforestation a key contributor to human-induced climate change (IPCC 2015). Today, almost 40% of the earth's non ice-covered terrestrial surface has been converted from natural lands to agriculture (Ellis *et al.* 2010). While the destruction and degradation of natural lands undoubtedly has an effect on carbon storage, empirical estimates of the carbon cost of land conversion have only recently been quantified. Despite the perception of deforestation as a clear-cut resulting in the absolute destruction of natural forests, most deforestation takes place piecemeal. Small forest patches are removed one at a time, and typically replaced by some form of agriculture. From a carbon point of view, this conversion may be less damaging than it would initially seem. Though the primary forest is gone, secondary forest may return, plantation forests may be implemented, forest fragments may remain, and logging in some areas may be selective. This results in a patchwork of different land cover types, with potentially varying carbon storage capabilities. Understanding how each of these types of forests functions in critical in modeling carbon dynamics on a landscape scale, and attempting to estimate how much they can contribute to climate change mitigation. As land conversion and climate change continue, efforts have increased to quantify carbon in modified landscapes.

When forests are cut, some patches are often left, surrounded by agricultural or other ecosystems. These forest fragments, despite a similar appearance, function fundamentally differently from continuous forest (Laurance and Bierregaard 1997). Biodiversity in forest fragments is frequently depleted, resulting in altered ecosystem functioning (Laurance *et al.* 2007). Given the high perimeter-to-area ratio, a large percentage of fragments are exposed to edge effects. These effects include altered microclimate and vegetation structure, which can affect the type of plants present and consequently the amount of carbon that can be stored (Murcia 1995; Laurance and Curran 2008). Osuri *et al.* (2014) examined how the physical parameters of vegetation are altered in fragments. Trees in these patches tend to be shorter and have less dense wood than trees in continuous forests. The mechanical pressure of winds at forest edges tends to remove the largest trees, resulting in carbon storage being reduced by 40%. The same pattern of limited carbon storage in fragments was found by Gilroy *et al.* (2014). The plant communities are different at forest edges, which generally means reduced carbon storage in tropical areas (Osuri *et al.* 2014), though not necessarily in temperate areas (Ziter *et al.* 2014). A large percentage of the area of a fragment is exposed to edge effects, so most of a fragment is reduced in its ability to store carbon. Thus, the reduction in carbon storage capacity is greater in magnitude than would be suggested by the reduction in forest area. Forest fragments cannot be considered the same as continuous forest—this should be factored into carbon modeling efforts.

In many areas, the primary forests have been removed completely, but secondary forest has grown back to replace it. Farms may be abandoned and timber cuts grow back, or ecological restoration efforts are made to implement new forests. Regrowth follows predictable patterns in most forests. The first species that come back are fast growing, typically light-wooded pioneers that live and die quickly. As forests age, these pioneers are replaced by much slower-growing, longer-lived species that have denser wood. The pioneer species can accumulate biomass rapidly, but they are often hollow, or have light wood, and cannot store a large total amount of carbon. The opposite is true of shade-tolerant species, which can take time to grow but ultimately contain large amounts of carbon. Given the different growth rates and wood traits, patterns of carbon accumulation change through both the life of the individual tree and the life of the forest. Shimamoto *et al.* (2014) have determined that fast-growing species accumulate more carbon than slower species during the first 40 years of life; after that, the slower species accumulate carbon more quickly. As forests age and slower species replace pioneer species, total carbon accumulation increases, and the value of the forest increases the older it gets. Restoration projects should include a mix of species for optimal carbon storage, long and short-term.

When natural forests are removed, they are often replaced with some form of artificial forest, such as plantations or agroforests (systems in which trees are incorporated into other crops, such as coffee). These artificial forests are most often very different from natural forests: plantations are often of a single, likely exotic species, and agroforests tend to have far fewer trees per hectare than natural forests (Moguel and Toledo 1999; Felton *et al.* 2010). Furthermore, species are often selected for traits such as use for construction or fruit, leading to non-random species sets. Agroforests run the gamut from a few trees in crop fields to dense, rustic plantations that closely resemble natural forests (Moguel and Toledo 1999). Tadesse *et al.* (2014) compared the carbon storage of various types of coffee agroforests compared to nearby large forest fragments. Small farm owners tend to select dense-wooded species that are good for charcoal and construction, and such farms can store up to two thirds of the carbon of the carbon in fragments. Larger commercial plantations had fewer trees but could still store about half the carbon of fragments. While this is a significant reduction from continuous forest, it is non-negligible and indicates the importance of agricultural forests. Plantation forests can also store significant levels of carbon (Ren *et al.* 2014). The species in the plantation determines exactly how much carbon can be stored, with native broad-leaf plantations storing as much as forests, while exotic soft woods may store only 25% as much. Plantation forests may be reduced in their capacity to store carbon, but implemented over wide scales, they have the potential to mitigate the effects of deforestation. Ren *et al.* found that while deforestation was rampant on Hainan Island in China, carbon storage capacity has been increasing due to the implementation of large-scale plantation forests in many areas. Artificial agricultural forests have significant potential as replacements for natural forest that maximize both carbon storage and economic productivity.

Degraded forests are also common in areas of high timber extraction. These forests occur when selective logging, hunting, or extraction of non-timber forest products is allowed (Burivalova *et al.* 2014). While the latter two tend not to be too destructive, selective timber removal can be highly invasive, removing large trees and destroying other vegetation. Following experimental tree removal plots, Sist *et al.* (2014) found significant reductions in carbon storage after selective logging. The effects of logging last beyond the initial extraction period as well; large trees not initially taken out in the logging suffer high mortality and morbidity, possibly due to damage sustained in the extraction. Small trees increase in biomass, but they are unable to compensate for the loss of large trees. Degraded forests may still be counted in geospatial surveys as regular forests, which fails to accurately represent spatial carbon distribution.

Several key factors determine the amount of carbon that can be stored in these forests: the size of the trees, density of the wood, and the number of trees per hectare. Managed forests and those under selective logging have fewer trees, and secondary forests have pioneer species with less dense wood. Large trees are also critical for carbon storage. Not only do larger trees simply hold more carbon, but also the rate of accumulation of carbon increases steadily with the size of the tree, with no indication of senescence (Stephenson *et al.* 2014). Large trees are the primary determinants of the carbon storage capacity of degraded forests (Sist *et al.* 2014) and as a result of their influence, forests steadily increase in carbon content as they age (Shimamoto *et al.* 2014). Agroforests and plantation forests are generally too young to be filled with large trees, and forest fragments are subjected to high mechanical pressure from wind and altered microclimates that increase mortality of large, tall trees. The forests that replace continuous primary forest are diverse, but united by reduced carbon capacity caused by a lack of large trees and altered communities.

Allometric and Structural Changes Reduce Carbon Storage in Forest Fragments

Contrary to the popular image of deforestation as a clear-cut resulting in the absolute destruction of forests, most deforestation in the tropics takes place piecemeal. As forest is logged or converted to agriculture, patches are often left standing, resulting in a fragmented system of forest patches in a mosaic composed primarily of agriculture. Forests in tropical areas are increasingly highly fragmented, which has significant impacts on biodiversity and environmental conditions within the fragments. However, little is known about the impact of fragmentation on the ability of the forest to store carbon. In an ever-more-fragmented, ever-more-carbon-saturated world, understanding how these remnant forests sequester carbon is critical. Osuri *et al.* (2014) examined the relationship between rainfall, fragmentation, and carbon storage in fragments and continuous forest in the Western Ghats of southern India. Using linear mixed models and regressions, they found that fragmented forests stored almost 40% less carbon than continuous forests, as a result of trees that were shorter, had less dense wood, and were shorter for a given trunk diameter. Fragmented forests also relied more on large trees to store carbon, while displaying signs of transitioning to a community of less-dense, lower-carbon species.

In order to determine how forest structures differed between fragments and continuous forest, Osuri *et al.* measured each tree with a diameter greater than 10 centimeters at chest height. They recorded diameter, canopy height, identified the species, and found the density of the wood. From these measurements they calculated the height-to-diameter ratio, the amount of biomass, and thus the amount of carbon present. Using multiple linear regressions, they compared how these factors varied between fragmented and continuous sites, with both fragmentation and mean annual precipitation as possible explanatory factors. They further used linear models to see how the height-to-diameter ratio changed within a species or a community between sites, and how carbon storage was distributed between size classes.

Though the continuous sites received more rainfall than the fragmented sites, precipitation was only primarily responsible for increased tree density in the continuous sites. Other differences were better explained by fragmentation: trees in fragments were 25% shorter, occupied 22% less space per hectare, were 6% less dense, and stored 36% less carbon per hectare. Fragment trees were also shorter for a given diameter, and large trees did the bulk of the carbon storage in the fragments, while in continuous forest storage was more evenly spread across size classes.

Trees in fragments are subject to different pressures than continuous forest trees: fewer trees around means more light, but it also means that high winds aren't stopped by the combined trees of the forest. In such conditions, growing tall is not only unnecessary, it can be damaging, as it makes the tree vulnerable to being blown-over by wind. Older trees present before fragmentation were still taller, but as the community ages, replacements can be expected to be shorter and smaller. In many fragmented areas, large old trees are lost quickly, brought down by wind and fire coming from open land like pasture adjacent to the fragment. Osuri *et al.* note that agroforestry systems like shade coffee plantations next to the fragments likely buffer them from these effects, preventing the early loss of old trees. However, while older trees prevent significant carbon losses in the present, as the community transitions to smaller, less dense individuals and species, high carbon losses may be inevitable. Ultimately, in order to mitigate this threat, active efforts will be needed to restore species composition and forest structure, and thus maintain carbon storage capacity.

Carbon Storage in Restored Forests is Species and Age Dependent

Deforestation in tropical rainforests is a significant and growing conservation concern, and for good reason: as well as harboring high levels of biodiversity, tropical forests are estimated to store 59% of global terrestrial carbon. The capacity of woody plants to store carbon, which constitutes 50% of their biomass, makes them an indispensible consideration in the effort to mitigate global climate change. Of course, forests can't store carbon if they don't exist. In the past 14 years alone, more than 100 million hectares of tropical forest have been lost—an area greater than Texas and Arizona combined. This continued destruction has prompted interest in the ability of ecological restoration—replanting forests—to provide ecosystem services such as carbon sequestration and biodiversity habitat. In attempting to rapidly revitalize damaged ecosystems, fast-growing, pioneer species with low wood density are often chosen to replant, though slower-growing, denser species may be required for long-term carbon storage and ecosystem health. To help resolve this question, Shimamoto *et al.* (2014) examined the biomass accumulation of ten tree species with different ages and growth patterns. By comparing measurements of fast and slow-growing trees in forests of different ages, they were able to determine carbon sequestration through analysis of covariance tests as well as linear and non-linear models. They found that in the first 35-40 years, fast-growing species accumulate the most carbon, but after 40 years, slow-growing species accumulate more carbon, and older forests overall sequester more carbon than young forests.

Shimamoto *et al.* selected 10 common species, 6 fast-growing and 4 slow-growing, and identified 6-19 individuals of each species. The sampled individuals were found in forests ranging from early restoration (7-11 years old) to late-stage secondary forests (20-60 years old). They measured each individual for height, diameter, wood density, and age (using small trunk cores), and used these measurements to calculate the amount of biomass and thus carbon present. They also measured canopy cover and vegetation density in circular plots around the trees, to determine if these had an effect on biomass. To compare the relationships between tree measurements, they used linear, polynomial, and logarithmic models, and determined which type fit best. They then used an analysis of covariance to compare the amount of carbon sequestered between fast-growing and slow-growing species. Patterns varied between species, but in general, older trees were taller and wider, with taller and denser trees accumulating the most biomass. This resulted in a variable but general pattern of biomass accumulation increasing with age. They also found that both fast and slow-growing species accumulated more biomass in areas with higher vegetation density, though only fast-growing species increased with higher canopy cover. Most importantly, they found that the type of growth pattern sequestering the most carbon changed over time. Initially, fast-growing species sequestered more carbon than slower-growing species, probably due to their rapid increases in height and diameter. This pattern persisted until approximately 38 years after replanting, when it switched, and the denser, slower-growing species began sequestering more carbon, a pattern that then persisted. The enhanced carbon capture capacity of slow-growth species caused older forests to accumulate significantly more carbon than younger forests, making them more valuable in efforts to alleviate climate change.

The selection of species for restoration efforts is difficult, and depends on the priorities of the specific site. Faster-growing species provide habitat for their species more quickly, increasing native biodiversity, arresting soil erosion, filtering water, and sequestering carbon with their rapid addition of biomass. However, slow-growth species are necessary to continue to accumulate benefits over the long term, and to provide a more complex assemblage that will further enhance species richness. Tree selection will differ for the needs of each specific project, but the models of Shimamoto *et al.* suggest that if the goal is effective long-term carbon accumulation, a mix of species will be most beneficial. Thus, restoration of native forests has the potential to provide numerous services and sequester significant amounts of carbon, and should be carefully considered as a strategy to mitigate global climate change.

Coffee Agroforests Can Store Significant Levels of Carbon

Over 40% of the world's terrestrial surface is covered by agricultural activities, and approximately half of that area is agroforestry. Agroforests, agricultural areas that are at least 10% covered by tree shade, run the gamut from areas with a few exotic species to structurally complex, highly diverse ecosystems that mimic natural forests. Woody vegetation biomass is approximately 50% carbon, so incorporating trees into agricultural areas significantly improves the ability of these systems to sequester and store carbon. Given that agroforests cover almost half a billion hectares, they may represent a significant and underestimated carbon sink. The amount of carbon that can be stored is determined by a variety of factors, including the number of trees and the density of their wood. The type of and number of the trees present in turn depends on the individual management of the farm. In Ethiopia, Tadesse et al. (2014) investigated how different management regimes affected the species of trees present and the amount of carbon stored, compared to natural forests. They measured the density and species of trees in smallholder coffee farms, state-owned plantations, and forest fragments, and used these measurements to determine the carbon storage capacity of each forest. They also interviewed farmers to see how and why species are selected for inclusion in plantations. Tadesse et al. found that agroforests can store 50-62% of the carbon that natural forests can store. They also found that farmers tended to prefer and harvest denser-wooded species, though less dense species were used for some limited purposes.

Tadesse et al. established plots in multiple forest fragments, smallholder farms (< 3 Ha), and state-owned plantations. They recorded the height, diameter, and density of all trees larger than 10cm diameter, as well as the species. From these measurements they calculated the aboveground biomass and amount of carbon stored in each type of system. They found that trees in plantations tended to have less dense wood and stored 50-62% of the carbon in natural forests. Natural forests had more samplings than agroforests, and while the wood density of smallholder farm species did not differ from natural forests, species in state-owned farms tended to be less dense. Carbon storage was higher in smallholder farms, where there were more trees (281 per hectare vs. 180 per hectare in state-owned plantations) and the trees tended to have denser wood.

The researchers also interviewed the owners of the small farms. To better understand why each species was selected, they asked about common uses and which species were preferred. They found that wood was commonly used for charcoal, construction, and fuel. Species were selected for a variety of factors, including straightness, easiness to split, flammability, and availability. Species with dense wood were preferred for construction and fuel, with only beehive construction using lighter wood. This resulted in dense-wooded species being heavily harvested in forest fragments and farms, though not in the better-protected state-owned plantations.

Both types of plantation, smallholder and state-owned, are capable of storing a significant amount of carbon. Smallholder plantations are more intensively used, with higher levels of harvesting than the protected state-owned plantations. However, they also have more trees and tend to have denser-wooded species, resulting in a carbon storage capacity that is almost 2/3 that of natural forest. Agroforestry may therefore be highly important in the effort to sequester carbon and mitigate climate change. These systems should be encouraged, but carbon storage considerations are likely to be beyond the concern of small-scale subsistence farmers. Instituting carbon credit programs would allow farmers to place a monetary value on the presence of living trees on their farms, and could provide the impetus for implementing more sustainable farming practices.

Plantation Forests Increase Carbon Storage on Hainan Island, Southern China

Forested landscapes in the tropics are often highly dynamic, with natural forest being replaced by a shifting mosaic of plantations, agriculture, pasture, and settlements, which are in turn occasionally replaced by ecological restoration. This produces landscapes that vary in their capacity to sequester and store carbon. In South East Asia, as in many parts of the tropics, natural forests are commonly converted to plantations of rubber, *Eucalyptus,* pine, and hardwood for timber. The carbon storage capacity of a forest depends on the species of trees, which vary in density and size. Consequently, the type of forest, artificial and natural, determines the carbon storage capacity of a landscape. On Hainan Island in Southern China, land cover has been continuously altered over the past century, with artificial plantation forests replacing natural rain forest. Ren *et al.* (2014) analyzed how carbon storage capacity differs between land use types and how the total quantity of stored carbon has changed through time. Using both remote sensing and forest inventory plots, they quantified the carbon stored in woody vegetation, understory, herbs, leaf litter, and soil in each type of land use. They found that carbon storage capacity is highest in natural forests, and while these forests have been reduced over the years, the proliferation of plantation forests has actually increased the total carbon storage of the island. Furthermore, they found that 75% of the forest's carbon was stored in soil rather than woody biomass, emphasizing the importance of maintaining soil communities.

Ren *et al.* used historical forestry survey data obtained from the Chinese Government to estimate the extent and type of forests from the 1940's onwards. To determine the capacity of these forests to store carbon, they established survey plots in forests of each type. They determined the dominant species, the average diameter and tree height, and collected litter and understory herbaceous plants. They also took soil cores to a depth of one meter. To determine the carbon storage capacity, the researchers calculated the forest stand volume in cubic meters, and then multiplied that by a species-specific conversion factor (similar to density) to determine the amount of biomass and hence carbon. Understory plants and leaf litter were dried and weighed, and soil samples were assayed for their organic matter content. These measures allowed Ren *et al.* to estimate the amount of carbon stored in an average hectare of forest of each type. Additionally, they estimated the number of trees present in urban areas and used this to estimate the carbon storage capacity of cities. They then used the calculated carbon storage capacity and spatial data detailing the extent of each forest type to determine the total amount of carbon stored on Hainan Island both historically and in the present day. The data revealed significant decreases in natural forest area throughout the 20th century, as well as the proliferation of plantation forests. Natural rainforests stored more carbon than any other forest type. However, while natural forests declined, total island-wide carbon sequestration increased from 1993 to 2008 as a result of the plantation forests. Understory plants and leaf litter were found to contain relatively little of the forests' carbon. Aboveground woody biomass constituted less than 23% of forest carbon; soil contained almost 75% of the carbon. Total carbon storage was approximately 280 teragrams, with an additional possible storage capacity of 77 teragrams if natural forest ecosystems were restored.

Carbon storage is, unsurprisingly, highest in undisturbed ecosystems with high biomass. However, plantation forests strike a favorable balance between economic profitability and carbon storage, and due to the former, are much more likely to be viewed as an acceptable form of land use. Restoring natural ecosystems would enhance carbon storage capacity, but due to the lack of financial incentives, natural rainforests are largely confined to remote mountainous regions. Regardless, the high levels of carbon on Hainan emphasize the ability of anthropogenic ecosystems to serve as tools to mitigate climate change. This research also shines a light on the importance of soil carbon. Three quarters of forest carbon is stored in the soil, so land use policies that prevent soil erosion and degradation should be encouraged. Ecological restoration of natural forests is best for carbon storage, but plantation forestry offers an attractive way to maintain soil and woody biomass while generating economic profit for private landholders.

Large Trees Drive Carbon Sequestration in Degraded Tropical Forests

Deforestation is responsible for 15% of human-caused carbon release and hence is a key driver of global climate change. However, less known is the role that degradation plays. Forests become degraded by persistent human use or through selective logging, decreasing biodiversity and potentially hindering ecological dynamics. In the Amazon basin, degradation may account for up to 25% of carbon emissions by land use. Selective logging commonly targets the largest trees, which by definition contain the most biomass and carbon. By removing these, logging often substantially reduces forest carbon stocks. Relatively little is known, however, about how this disturbance affects biomass dynamics among size classes at a tree stand level. Sist *et al.* (2014) address this deficit by following biomass changes among trees of various size classes through 8 years after selective logging. They surveyed 18 experimental plots every two years, collecting data on biomass changes within individual trunk diameter categories and on mortality or morbidity in each category. They found that while small trees increased in biomass, large trees are the key drivers of ecosystem carbon storage. Large trees account for close to half of total carbon storage and experienced high post-logging mortality, which caused significant carbon losses. In order to compensate for this, the authors conclude that logging intensity may need to be reduced and a maximum diameter cutting limit should be adopted.

In order to determine the biomass changes taking place, Sist *et al.* established 18 one-hectare experimental plots in a forest designated for logging. They performed an initial survey of the plots, measuring tree density, basal area, and tree diameter. From the measurements of diameter and previously recorded density, they estimated the amount of biomass and hence the amount of carbon. The plots were then logged. Three months after logging, they were resurveyed to determine mortality and injury to all trees present. For the next 8 years, the plots were resurveyed approximately every two years, and the biomass calculations were performed again. Sist *et al.* found that prior to logging, large trees represented less than 10% of the tree density, but stored 49% of the carbon. Following logging, which targeted large trees, the plots experienced a significant loss of carbon. In the eight years after the logging event, biomass changes differed between size classes. Smaller trees tended to gain in biomass, and increased their rate of gain. However, medium and large trees tended to lose biomass as a result of residual mortality due to injury from the logging. This mortality in the larger size classes balanced biomass gains by small trees and led to a relatively constant overall biomass balance. The researchers also found that small trees sequester about 12 kilograms of carbon per year, while medium trees sequester 30 kilograms and large trees sequester 53 kilograms, more than 4 times as much as small trees. Sist *et al.* estimated that it would take more than 50 years for the forest to recover its biomass, but if logging intensity were reduced from 6 to 3 trees per hectare, recovery would only take 15 years. This recovery would be aided by preventing the cutting of the largest trees.

This research demonstrates the significant impact of large trees on carbon dynamics in all forests, and the downsides of removing them via selective logging. They are often selected for logging due to raw timber volume; however, many have structural defects that limit the amount of wood that can be recovered. Thus, leaving some of the largest trees may be less costly to loggers than timber volume would suggest. Reducing the logging intensity and leaving the largest trees remaining would significantly improve the ability of the stand to regenerate the lost biomass quickly, though this would present a cost to loggers. To help defray these costs, carbon credits could be issued that pay to keep the trees alive and the carbon sequestered. In order to make the prices competitive, the credits would have to be issued for at least $6.50 per ton of carbon. Damage to the stand overall could also be reduced by selecting only trees that show an increased risk of mortality, easily assessed via crown damage, liana infestation, etc. Better training and supervision of logging crews could in this way greatly reduce the damage done by selective logging to forest carbon storage.

Large Forest Blocks are Essential for Biodiversity Protection and Carbon Storage

Habitat loss is the primary threat to the survival of most tropical biodiversity. Typically, this habitat loss is driven by deforestation for agricultural use. However, deforested landscapes are rarely homogenous fields with low diversity; most often, forest fragments are left embedded in a matrix of varying types of agriculture, from open field monocultures, to pastures and forest-mimicking shaded plantations. The process of fragmentation has a significant negative effect on the biodiversity present in the area; however, fragments are often able to support a variety of species, as are some types of agriculture, such as agroforestry. Less is known about the capacity of such landscapes to sequester and store carbon. What little has been done has focused on carbon in agroforestry systems, with promising, though mixed, results. These conflicting results have led to a debate among conservationists about the best way to protect natural ecosystems. Some argue that agriculture should be intensified on smaller amounts of land, allowing large, continuous blocks of land to be left untouched (land-sparing). Others advocate incorporating natural elements (trees, shrubs, etc.) into agriculture, to make it more hospitable to biodiversity (land-sharing). However, most arguments one way or another have been theoretical, with little supporting empirical evidence. What evidence does exist is typically spatially limited, with few landscape-level studies being performed. To help resolve this debate, Gilroy *et al.* (2014) present the first integrated study of both multi-taxa biodiversity and ecosystem carbon storage capacity. They quantified carbon contained in forest fragments compared to the same size plots in continuous forest, and surveyed birds and dung beetles in each habitat type. Based on the data collected, they used land-allocation models to predict how biodiversity and carbon would respond under various forest cover scenarios. They found that the greatest amount of biodiversity and carbon could be protected in landscapes in which agriculture was intensified on a smaller amount of land, with large continuous forest blocks left unharmed.

To determine the amount of carbon present in the landscape, Gilroy *et al.* established plots in forest fragments and advanced secondary forest, as well as control plots in primary forest. They measured the diameter and density of the trees in each plot, as well as the mass of leaf litter. They also estimated root biomass through root-to-shoot ratio data previously established in the literature. Through allometric measurements in the literature, they used the density and diameter to determine the amount of carbon in each type of ecosystem. Gilroy *et al.* also established larger plots to sample the biodiversity. They performed point counts at three points in each plot to visually assess the bird species present. They also placed baited pitfall traps to capture dung beetles, which are considered an indicator taxon for other biological groups. The variance in carbon stocks was assessed using analysis of covariance tests, and biodiversity was examined using multi-dimensional scaling algorithms and logit-link functions, which assess the community structure and probability of occurrence at each site. The researchers then used these data to construct a set of landscape models. Holding production area constant, they examined how various landscape configurations would affect carbon and biodiversity present. The land-sparing scenarios involved continuous blocks of primary forest or blocks of secondary forest. The land-sharing scenarios they used involved productive areas sprinkled with forest fragments. After running the models many times, they determined which configurations resulted in the highest values. They found that carbon storage was significantly lower in forest fragments compared to primary forest. Advanced secondary forest had lower carbon stocks than the primary forest, but contained more than forest fragments. Biodiversity occurrence followed the same pattern as carbon storage. The models found that land-sparing scenarios resulted in the highest carbon storage and biodiversity protection, with little difference between primary forest and secondary forest blocks.

Land-sparing versus land-sharing has been a hotly debated topic in the conservationist community. However, little empirical evidence has been offered to support one side or the other. Gilroy *et al.* produced one of the first landscape-level models calibrated with field data to examine broad patterns in biodiversity and ecosystem services. The results indicate that forest fragments are of relatively little value in productive systems. Rather than attempt to maintain fragments, conservationists should look to improve future outcomes by promoting increased yields in intensified plots, while leaving large continuous blocks of forest untouched. Such systems can be successful both

in previously undisturbed systems in which new agriculture is form-
ing, and in old agricultural landscapes where forest blocks will be
composed of secondary regrowth.

Carbon Storage Increases Continuously as Trees Grow

Though it has been assumed that the rate of carbon accumulation declines with the age of an individual tree, little empirical evidence has been produced to support this assumption. Understanding
how carbon storage capacity changes throughout the life of the tree is
important in modeling carbon dynamics in forests, which can be used
to determine how forests will contribute to climate change mitigation
plans. Net primary productivity is well known to decline in even-aged
forests, as does mass gain per unit leaf area. However, few forests are
completely even-aged, and many are subjected to selective logging
that removes the largest trees. Proper modeling of the amount of carbon lost through this logging can be used to more accurately price
carbon credits for the preservation of natural forests, aiding efforts to
keep them intact. In order to determine how carbon storage rates
change with tree age, Stephenson *et al.* (2014) collected data from
long-term monitoring plots in tropical and temperate areas across the
globe. By measuring the diameter of each tree and using allometric
equations, the researchers determined how much carbon was being
stored over time. They found that while stand productivity declined
with age, individual tree carbon gain rate increased, with no signs of
declines at any age.

Stephenson *et al.* used data from long-term ecological monitoring plots in forests on every continent. Forests in each plot are surveyed approximately every five to ten years, with measurements re-
taken on each individual tree. The diameter of each tree was measured, and combined with allometric equations in the literature to determine the amount of carbon present in each tree. Measurements
were compared for individual trees through time to generate the
change in growth rate with age. Models were constructed to determine if the shape of the relationship was increasing, decreasing, or
parabolic, and the Akaike Information Criterion was used to deter-

mine which model matched the data most effectively. They also tested for bias caused by cross-plot measurements or differing allometry, but found no systematic errors. The data included measurements on 403 species in both tropical and temperate areas, and found that rate of mass gain increased with tree mass in 87% of the species examined. The authors found no evidence of a decline even in the largest trees, with the largest trees potentially adding more than half a metric ton of biomass per year.

There is no evidence that large trees decline in their ability to add mass and sequester carbon. Thus, older trees do not act merely as passive reservoirs of carbon, but actively sequester significant levels of carbon. This contrasts with previously noted declines in stand productivity, which can be explained by reductions in tree density. This research also indicates the value of large trees in carbon capture schemes. Large trees are often removed during the selective logging process, potentially severely reducing the amount of carbon storage in the forest. This can help inform carbon credit values, which are proposed to reward farmers who incorporate trees into agroforestry systems. Larger, older trees are more valuable, and consequently carbon credits should increase on older farms. The results also indicate that reductions in growth rate may not be associated with senescence, as previously assumed.

Community Composition is Different at Forest Edges, but Carbon Storage Remains the Same

Forest fragmentation is one of the leading ways that humans alter natural habitat. Forests are frequently fragmented as land is cleared piecemeal for the expansion of agriculture, logging, and human settlement. Often, rather than clearing an entire forest, fragments of forest are left embedded in a matrix of agricultural and other habitats. As an increasing percentage of the world's forests are fragmented, it is crucial to understand how forest fragments function. Fragments are subject to a variety of influences, most notably edge effects. Edge effects occur at the edges of two habitats, and include altered microclimate, reduced biodiversity, and vegetation changes.

These edge effects can bring about altered species communities, which in turn could affect the amount of carbon that can be sequestered near forest edges. As forest fragmentation continues, a greater percentage of forest will be exposed to edge effects, potentially inhibiting forests' ability to act as carbon sinks. To understand these effects, Ziter *et al.* (2014) examined how tree species composition and carbon storage capacity change with proximity to forest edge in large and small fragments. Using tree measurements and allometric data in the literature, they determined how much carbon was stored, and which species were present. Using linear mixed models and multidimensional scaling, they found that community composition shifts with proximity to the forest edge. Despite this shift, however, carbon storage did not decrease closer to the edge.

Ziter *et al.* identified six each of small connected, small isolated, large connected, and large isolated secondary forest fragments embedded in an agricultural matrix in Quebec, Canada. Two 100 meter transects were established in each fragment, with five 10x10 meter plots per transect. In each plot, the researchers recorded the species and diameter at breast height of all stems. Using allometric equations from the literature, they determined how much carbon was stored in each plot. Linear mixed models were used to assess how proximity to the forest edge affected both species richness and carbon storage. To test community composition, they used non-metric multidimensional scaling, which compares the similarity of communities. They found no effect of any variable on carbon storage; carbon stocks remained constant over the gradient from forest edge to forest interior. However, vegetation structure and species composition was altered at the forest edge. Stem density and species richness were higher at the edge, and included more shade-intolerant species.

The lack of a decline in carbon storage at the forest edge contrasts with previous studies. In tropical areas, large trees are responsible for the majority of carbon storage in a forest, but are often removed from edges by wind pressure and altered microclimate. While a similar pattern was observed in this study, increased vegetation density at the edge compensated for the reduction in large trees. In some fragments, large trees were still preserved at the edge, which is likely the result of human management for large sugar maples. This study indicates important differences between the reactions of temperate and tropical forests to fragmentation. While tropical forests are reduced in their capacity to sequester carbon, temperate forests appear to maintain full storage capacity. Thus, even small and irregularly shaped fragments may remain valuable for carbon storage in human-dominated landscapes.

Conclusions

Despite the rapid and ongoing destruction of primary forest across the globe, there is reason to be optimistic about the ability of modified ecosystems to store carbon. Deforestation is better thought of as land conversion to a diverse set of human-dominated ecosystems, many of which have the ability to sequester and store carbon. Without exception, modified ecosystems store less carbon than continuous forests; while the amount of carbon that can be stored varies with multiple factors, it is typically 30% to 60% of that stored in primary forests. Carbon storage capacity increases with the age of trees, and consequently large trees are the primary drivers of carbon dynamics. Large trees are typically removed in modified landscapes, through the action of selective logging, agricultural turnover, and edge effects, resulting in smaller carbon stocks. Still, on a broad scale the use of plantation and agroforests can maintain regional carbon storage. Carbon credit programs could be useful in encouraging these types of land use, especially for small farmers who use diverse, dense species. Deforestation, forest degradation, and land conversion are still significant threats to biodiversity and carbon storage, but the recent research suggests the situation may be more hopeful than we thought.

References Cited

Burivalova, Z., Sekercioglu, C.H., Koh, L.P., 2014. Thresholds of Logging Intensity to Maintain Tropical Forest Biodiversity. Curr. Biol. 24, 1893–1898. doi:10.1016/j.cub.2014.06.065

Ellis, E.C., Goldewijk, K.K., Siebert, S., Lightman, D., Ramankutty, N., 2010. Anthropogenic transformation of the biomes, 1700 to 2000. Glob. Ecol. Biogeogr. 19, 589–606. doi:10.1111/j.1466-8238.2010.00540.x

Felton, A., Knight, E., Wood, J., Zammit, C., Lindenmayer, D., 2010. A meta-analysis of fauna and flora species richness and abundance in plantations and pasture lands. Biol. Conserv. 143, 545–554. doi:10.1016/j.biocon.2009.11.030

Gilroy, J.J., Woodcock, P., Edwards, F.A., Wheeler, C., Medina Uribe, C.A., Haugaasen, T., Edwards, D.P., 2014. Optimizing carbon storage and biodiversity protection in tropical agricultural landscapes. Glob. Change Biol. 20, 2162–2172. doi:10.1111/gcb.12482

Intergovernmental Panel on Climate Change, 2015. Climate Change 2014 Synthesis Report. Cambridge University Press.

Laurance, W.F., Camargo, J.L.C., Luizao, R.C.C., Laurance, S.G., Pimm, S.L., Bruna, E.M., Stouffer, P.C., Williamson, G.B., Benitez-Malvido, J., Vasconcelos, H.L., Van Houtan, K.S., Zartman, C.E., Boyle, S.A., Didham, R.K., Andrade, A., Lovejoy, T.E., 2011. The fate of Amazonian forest fragments: A 32-year investigation. Biol. Conserv. 144, 56–67. doi:10.1016/j.biocon.2010.09.021

Laurance, W.F., Curran, T.J., 2008. Impacts of wind disturbance on fragmented tropical forests: A review and synthesis. Austral Ecol. 33, 399–408. doi:10.1111/j.1442-9993.2008.01895.x

Moguel, P., Toledo, V.M., 1999. Biodiversity conservation in traditional coffee systems of Mexico. Conserv. Biol. 13, 11–21. doi:10.1046/j.1523-1739.1999.97153.x

Murcia, C., 1995. Edge Effects in Fragmented Forests - Implications for Conservation. Trends Ecol. Evol. 10, 58–62. doi:10.1016/S0169-5347(00)88977-6

Osuri, A.M., Kumar, V.S., Sankaran, M., 2014. Altered stand structure and tree allometry reduce carbon storage in evergreen forest fragments in India's Western Ghats. For. Ecol. Manage. 329, 375–383. doi:10.1016/j.foreco.2014.01.039

Ren, H., Li, L., Liu, Q., Wang, X., Li, Y., Hui, D., Jian, S., Wang, J., Yang, H., Lu, H., Zhou, G., Tang, X., Zhang, Q., Wang, D., Yuan, L., Chen, X., 2014. Spatial and Temporal Patterns of Carbon Storage in Forest Ecosystems on Hainan Island, Southern China. PLoS One 9, e108163. doi:10.1371/journal.pone.0108163

Shimamoto, C.Y., Botosso, P.C., Marques, M.C.M., 2014. How much carbon is sequestered during the restoration of tropical forests? Estimates from tree species in the Brazilian Atlantic forest. For. Ecol. Manage. 329, 1–9. doi:10.1016/j.foreco.2014.06.002

Sist, P., Mazzei, L., Blanc, L., Rutishauser, E., 2014. Large trees as key elements of carbon storage and dynamics after selective logging in the Eastern Amazon. For. Ecol. Manage. 318, 103–109. doi:10.1016/j.foreco.2014.01.005

Stephenson, N.L., Das, A.J., Condit, R., Russo, S.E., Baker, P.J., Beckman, N.G., Coomes, D.A., Lines, E.R., Morris, W.K., Rueger, N., Alvarez, E., Blundo, C., Bunyavejchewin, S., Chuyong, G., Davies, S.J., Duque, A., Ewango, C.N., Flores, O., Franklin, J.F., Grau, H.R., Hao, Z., Harmon, M.E., Hubbell, S.P., Kenfack, D., Lin, Y., Makana, J.-R., Malizia, A., Malizia, L.R., Pabst, R.J., Pongpattananurak, N., Su, S.-H., Sun, I.-F., Tan, S., Thomas, D., van Mantgem, P.J., Wang, X., Wiser, S.K., Zavala, M.A., 2014. Rate of tree carbon accumulation increases continuously with tree size. Nature 507, 90–+. doi:10.1038/nature12914

Tadesse, G., Zavaleta, E., Shennan, C., 2014. Effects of land-use changes on woody species distribution and above-ground carbon storage of forest-coffee systems. Agric. Ecosyst. Environ. 197, 21–30. doi:10.1016/j.agee.2014.07.008

Ziter, C., Bennett, E.M., Gonzalez, A., 2014. Temperate forest fragments maintain aboveground carbon stocks out to the forest edge despite changes in community composition. Oecologia 176, 893–902. doi:10.1007/s00442-014-3061-0

Effects of Drought and Fire in Amazonian Rainforests

Maithili Joshi

The relationship between fire-induced tree mortality and extreme weather remain poorly understood because it is restricted to post-fire observations of tree mortality. Studies done on the effects of forest fires and biodiversity remain understood on the patch scale, and do not consider the effects of fire on vegetation dynamics and structure. In the southeast Amazon forest, scientists established a large scale, and long term prescribed forest fire experiment in a transitional forest. Primarily, trying to determine if there are weather, and fuel, related thresholds in fire behavior associated with high levels of fire-induced tree mortality across two different fire regimes, and secondarily, what the effects of an intense forest fire are on forest structure, flammability, and aboveground live carbon stock.

Two experiments were conducted. First, three 50 hectare lots were established in a region with no signs of recent fires. The three lots were an unburned control, a lot burned every three years (2004–7), and a plot burned annually (2004–9). Second, regional analysis of weather and fire scars was studied in which burn scars in forested areas were mapped based on Landsat images.

In the 2007 drought, precipitation across the Xingu region was lower than in any other year recorded. This resulted in high-intensity fires. The difference in fire treatment is associated with fire spread rate given that "fuel composition". More leaf litter, and other things on the ground contributed to an increase in forest fires. These fires did not self-extinguish at night as they often do in non-drought years. This is attributed to high vapor pressure deficit, low litter moisture content, and high fuel loads.

In 2007 there was a sudden spike in tree mortality compared to other years. Post-fire mortality was pronounced along the forest edge treatment burned every three years, which mimics the rate of mortality in the region. This suggests that transitional forests are more resistant to low-intensity fires than wetter Amazonian forests because mortality rates in wetter forests tend to be considerably higher.

These results led to two drivers of fire intensity and fire induced tree mortality. First, 2007 was dryer and warmer, with higher fine fuel loads. The dry and warm weather conditions increased fuel loads, a likely cause of increased fire intensity. Second, fuel loads and mortality rates were higher along the forest edge, suggesting a difference in fire-induced tree mortality due to fine fuel loads.

Above ground biomass reduced significantly because of elevated fire-induced tree mortality. This led to reduced leaf area index causing more light to permeate, leading to more dryness in the understory. Along the edges of burned plots, grasses invaded, increasing fire intensity because grasses accumulated more fine fuel close to the ground than the trees they replaced. The grass invasion suggests that high intensity fires could promote abrupt fire-mediated transitions from forest to new stable states. These transitions are more likely to occur in fragmented forest areas, where disturbances are frequent and dry seasons are prolonged.

Another factor affecting tree mortality is deforestation, or other human disturbances. These affect mortality by reducing canopy cover and evapotranspiration. An increase in the average dry season and land-surface temperatures causes forests to become dryer and more prone to forest fires. Deforestation also creates a larger perimeter of forest fragments. Finally, tree mortality associated with logging, fire, drought, or edge effects contributes to coarse fuel loads for multiple years with twigs, branches, and standing dead trees decaying and falling to the ground. These forest fires also contribute to the climate of the Amazon, most forest burned in 2007 experienced another drought later, signifying that previous forest fires leave these areas more vulnerable to future ones.

Effects of Ant-Fruit Interactions on Deforestation

Biodiversity within an ecosystem has mutualistic and symbiotic relationships within that environment. The results of deforestation can be dramatic to these relationships, especially in cases with frugivores. The relationships between frugivores and fallen fruit are what help disperse seeds across the forest floor, which also helps the process of germination. In this study, Bieber *et al. (2014)* analyzed the mutualistic interactions between ants and fallen fruit in São Paulo State, SE Brazil. The scientists were examining the difference in interactions between disturbed and undisturbed forests. They compared the richness of ants at each fruit, species density per station, frequency of specific ant groups, frequency of fruit and pulp removal, and distance of fruit removal. The study was conducted using four disturbed forests, and four undisturbed forest areas. In these areas, there were thirty sampling stations with synthetic fruit placed 10 m apart from each other to ensure independent discoveries. The fruit were placed on a white sheet of paper within a wire cage to ensure that vertebrates did not access the fruit at each sampling station. The synthetic fruit are used in order to have a sufficient number for experimentation; these are similar to real fallen fruit because of its lipid-rich pulp for which ants show a high preference. The ants were tabulated by species type and were collected after 5 individuals visited the fruit at specific times of day. The researchers stopped recording at a fruit after it was 75% of its pulp was gone, or after 22 hours of exposure. The scientists found a total of 51 ant species attracted to the synthetic fruit, with ant richness ranging from 16 to 24 species per forest site. However, species richness was different between undisturbed and disturbed forest areas. They found, in general, that fragmented forests have lower overall species richness. They also found that large ponerines, *Pheidole* and *Solenopsis,* had the greatest number of species in both habitat types, while the species *Pachycondyla striata, Odontomachus chelifer,* and *Pheidole* were the most frequent removers of the lipid-rich pulp.

The authors observed that the large ponerines and large *Pheidole* were most likely to disperse the seeds, meaning this particular species is integral in maintaining this mutualistic relationship. They also discovered that ants found the fruit faster in undisturbed sites than in disturbed ones, leading to greater fruit removal at the undisturbed sites, although fruit removal was more variable in disturbed forests. The number completely cleaned fruit did not differ between the two forest sites. These results suggest there is a negative impact on ant-fruit interaction as a result of deforestation. Overall, undisturbed forests had denser species richness and assemblage of ant species interacting with the synthetic fruit than did fragmented forests. Undisturbed areas exhibited a much higher rate of ants finding the fruit in general, although ants still found the fruit in disturbed forests but not as quickly as they were found at undisturbed sites. This suggests that the fallen fruit in disturbed areas are less likely to be dispersed or germinate because there is a smaller likelihood of large ant species attending to them. There was also a significant difference in ant species composition between the two forest types, which is consistent with previous findings that point to lower species density in fragmented forests than in untouched forests.

Analyzing the Vulnerability of Rainforest Birds to Deforestation

In South East Queensland, Australia Pavlacky *et al.* (2014) conducted a study on the vulnerability of birds, rainforest ecosystems, and the biological impacts in response to deforestation in local and regional areas. The central idea is the to investigate the life history and forest structure to rank the vulnerability of avian species, while also looking at species loss along different kinds of forest structure and landscape change. The objectives are evaluating the effects of life history traits on the patch occupancy and vulnerability of rainforest birds, determining the relative effects of stand, landscape, and patch structure on species richness, and evaluating the relative contributions of deforestation and fragmentation to species richness.

Packlacky *et al.* categorized 46 different rainforest patches into three different rainforest types: upland notophyll vine forest, lowland notophyll vine forest, and araucarian notophyll-microhyll vine forest. They counted bird species along transects, recording time of day, date, and year. Life history traits of the birds were used to hypothesize the influence of extinction risk; These tasks included extent of occurrence, body mass, dispersal strategy, clutch size, feeding behavior, and population density. The scale of habitat, defined as "the spatial extent of ecological process", was also analyzed.

Land cover data, statistical analysis and model developments were used to study local landscapes, emigration and immigration patterns, and rainforest cover types. It was found that species with low population size were more vulnerable to fragmentation and deforestation than abundant species. When looking at behavioral specialization, there are high rates of extinction for species that live in terrestrial, understory, and canopy regions. Dispersal is another important factor which helped identify that sedentary species exhibited higher vulnerability than migratory or nomadic movements. Lastly, body mass analyses suggested that smaller species are more vulnerable to deforestation than are larger species. In looking at species richness and landscape structure, it was concluded that supporting avian communities is related to rainforest conditions at the stand sale, forest composition at landscape scale, and rainforest configuration at patch scale. There is evidence that bird assemblage responded to stand basal area, meaning that maintaining stand basal area is essential in minimizing the impacts of invasive plant species. This suggests that declining stand basal area and the degradation of forest structure could be one explanation of reduced species of birds.

There is a positive effect of maintaining forest cover in the surrounding landscape in order to maintain landscape connectivity, increase species richness, and increase occupancy of these birds. Further, it was found that vegetation composition could also improve habitat connectivity. As a result, it is important to improve stand condition in order to keep high species richness.

Finally, in looking at the effects of deforestation and fragmentation, they concluded that fragmentation exacerbates the effects of deforestation. In conclusion, maintaining high species richness, and biodiversity will be achieved by improving the structure of the rainforest.

Impact of Landscape Composition and Configuration on Bird Species in Lacandona Rainforest

Rainforests are rapidly being converted to human modified tropical landscapes (HMTLs) due to accelerated growth and demand for agricultural products. This study distinguishes between composition and configuration of tropical ecosystems to understand species responses to land changes due to HMTL's and improve management and conservation strategies in the tropics. Specifically, examining the relative effects of landscape composition and configuration on alpha and beta diversity of birds within old-growth forests in a fragmented biodiversity hotspot in the Lacandona rainforest in Mexico. The authors predicted that both alpha and beta diversity will have stronger associations with landscape composition than with landscape configuration, especially with specialist species. The percentage of forest cover in old growth forests and compared it to secondary forests was studied, and finally the composition of biodiversity within these two matrices was also examined.

In landscape studies, response variables were evaluated within old-growth forest patches and landscape variables were measured in a particular radius from the center of the patches in 20 sites. These sites were classified into six different types of landscape cover. A program was used to characterize spatial configuration of the landscapes surrounding the sites, and bigger sizes were selected to include home ranges of several bird populations.

The results showed that landscape composition and configuration affect the diversity of rainforest birds in the Lacandona region. It was found that old-growth forest cover showed the strongest association with diversity of forest specialist species. Second, forest edge density was positively related to the diversity of forest specialist species, whereas the percentage of secondary forest in the matrix tended to be negatively related to alpha-diversity of both forest specialist and generalist species. Third, the strength of all associations was higher for forest specialist than for habitat generalist species.

In conclusion, bird diversity was more strongly influenced by local-scale and landscape patterns, this can be associated with species diversity that was evaluated at the patch level. However, it was found that forest cover loss on alpha-diversity of specialist birds was consistent among all patch scales. This suggests that current land-use changes throughout the region will continue to have a detrimental effect on species in the near future; However, habitat generalists seem to be favored by deforestation.

Quantifying the Implications Protected Area Downgrading, Downsizing, and Degazettement (PADDD) for REDD+ Policies

REDD+ policies address deforestation and degradation of protected forests. It is believed their implementation causes perverse affects leading to illegal activities, downgrading, downsizing, and degazettement (PADDD). This phenomenon challenges the idea of permanence of protected areas. The study was conducted in the Democratic Republic of the Congo (DRC), Malaysia, and Peru because of its extraordinary biodiversity. Forrest *et al.* (2014) aimed to quantify the implications of PADDD for REDD+ polices. First, a database that consisted of information on PADDD events since 1990 until 2011 was created. This included protected area name, location and area affected, type, and year. Protected area legislation in these three countries and administrative journals in DRC were reviewed, and also digitized historic maps of PADDD events from government sources. Second the amounts and rates of deforestation and carbon loss within PADDDed lands in peninsular Malaysia and Peru were assessed and compared to unprotected forests. The events only included downsizings and degazettement. Analyses were based on four events: areas affected by PADDD, the current protected areas in Peru and peninsular Malaysia, above-ground biomass for the year 2007, and forest cover change in Peru to determine total forest cover lost, total forest carbon lost, and annual percent of original forest cover and forest carbon lost by land tenure class. The impact of protection and PADDD (in Peninsular Malaysia and Peru) on above-ground carbon stocks was estimated for the year 2010. This was compared to areas that have never experienced protection. For this experiment, three scenarios were examined: 2000–2010 observed rates of forest carbon loss proceed to 2100, 2000–2010 rates of deforestation, and all standing forests converted to non-forests by 2100. The economic net present value of forest carbon under three emissions and 3 carbon price scenarios was examined for Peninsular Malaysia and Peru and projected carbon value annually from 2010 to 2100. Additionally, the value of carbon lost between the years 2000 and 2010 was calculated. Finally, regression analysis of deforestation was used to test the hypothesis that PADDD is a significant predictor of deforestation in Peninsular Malaysia and Peru while accounting for biophysical characteristics and accessibility. The results showed that in DRC, there were 39 PADDD events; Malaysia experienced at least 121 PADDD events since 1900, as well as 110 in Peninsular Malaysia. The regression models suggested that PADDD is a significant predictor of forest loss. In Peru, there were 14 PADDD events that occurred in Peruvian protected areas. Thirteen occurred in 1996, resulting in 17% of the historic protected area system being permanently destroyed. Regression analysis also suggested that PADDD is a significant predictor of canopy cover change. The results show that PADDD has substantial implications for biodiversity conservation and ecosystem services. It can result in dramatically higher carbon emissions and deforestation rates.

To address these issues, the authors suggest robust social and environmental safeguards, comprehensive carbon accounting, rigorous monitoring, reporting, and verification systems and mechanisms for periodic policy reforms are crucial for climate policies to reach their desired goals, while minimizing consequences and outcomes.

The True Effect of Emission Reductions Resulting from Indonesia's Moratorium 2011

The impacts of deforestation are clear; the destruction of our forests results in increasing greenhouse gas emissions into the atmosphere. In an attempt to curb this, in 2011, Indonesia implemented a nationwide moratorium on peat lands, such as oil palm. Although this enactment is regarded as widely important, there are many questions raised about the moratorium. For example: how effective is the moratorium? Does it cover enough land, or the "right" lands? By how much did deforestation occur with the moratorium in comparison to prior to the concessions being granted from the years 2000 to 2010, before the moratorium was enacted? And, finally, how much lower would Indonesia's carbon emissions be if there had been no new concessions granted on primary forests and peat lands in those years prior? The answers to these questions are crucial for our understanding in carbon emissions and the effectiveness of the ban on deforestation. By estimating carbon emissions before the moratorium's enactment, using an empirical approach, and utilizing previously collected data to estimate carbon emissions, the evidence shows that it did not cover enough forest that is thought to hold large quantities of carbon. Simply put, the effect of the moratorium itself is insufficient. It is still agreed that without the law, the rate of deforestation would have occurred at a much faster rate had it not been introduced at all. However, the effectiveness of it would have a greater impact on reducing carbon emissions if policy been implemented on timber concessions as well as for oil palm because timber, on average, had a higher carbon density than oil palm concessions. The effect of a moratorium that covered a larger scope of the forest will drive carbon emissions down to reach the target the Indonesian government intended. Although these findings suggest that the suspension on deforestation did reduce emissions, there are other alternatives that have a greater impact, such as carbon-pricing mechanisms as a more effective way because of the economic incentives, thus making it easier to implement in general.

Implications of Land-Cover Changes and Fragmentation For Biodiversity Conservation

Deforestation can have substantial impacts on the vast biodiversity within tropical rainforests. In Hainan, China, importance is placed on trying to protect the habitat and biodiversity in the natural forests, specifically in the Changhua watershed. The Changhua watershed is an important area for China because it has been identified as the "center of endemism for plants and birds", so conserving this area is particularly important for maintaining biodiversity. In the last few years, biodiversity has been threatened by new rubber and pulp plantations causing forest fragmentation and larger patch distances. In this study, Zhai *et al.* looked at the implications of deforestation on biodiversity, especially of endemic species and the ecosystems surrounding using land cover data. Trying directly to quantify biodiversity in a landscape turned out to be hard, and furthermore difficult to interpret both land-cover data and fragmentation. The scientists analyzed land-cover satellite images for the years 1988, 1995, and 2005 and then compared natural forests and plantations in the Changhua watershed area. They used six types of land cover, including: natural forests, natural shrubs and grasslands, tropical crops, rubber plantations, pulp plantations, and open areas. In the years 1988 –1995, natural forests had increased in area, and only natural shrubs and grasslands decreased. Only a small number of natural forests were converted into other land covers. In the next decade, there was a drastic change natural forests, shrubs, and grasslands land cover, while pulp plantations and rubber plantations showed the highest increase. Overall fragmentation continued to increase, while habitat quality decreased. At the same time, there was an increase of fragmentation. All these results lead to the conclusion that that during the most recent time period of the study, 2005, land-cover changes resulted in loss, isolation, and fragmentation of protected areas. The implications of this fragmentation and isolation are substantial. First, it will decrease genetic diversity in flora and fauna, and affect the survival and reproduction of many endemic and endangered species. There will also be changes in community structure, composition, and species richness. With the beginning of agro-forestry there are implications that the disturbance to many of the natural forests will lead to invasive species that are detrimental to the survival of native species. The displacement of natural ecosystems caused a negative effect on the local biodiversity, meaning that these plantations will have an impact on the endemic species that are so important to the Changhua watershed. Further, there are several implications on biodiversity conservation as a result of forest fragmentation. The small fragments were not adequately able to support enough species, especially endemic species. Moreover, patch numbers, which are an ecological effect of fragmentation, increased, causing longer distances between patches, hurting genetic diversity because of the habitat isolation. This barrier is a concern to biodiversity conservation because of certain species, especially plants, that are used in order to track climate change. Without these species, it will be increasingly difficult to monitor changes. Structural damage to vegetation is another important factor in this discussion because it disrupts to the forest floor, soil layers, nutrient cycling, and decomposition, completely altering the composition of the natural forests, making it difficult for the original species to re-grow in these areas. Habitat fragmentation, loss, and isolation have detrimental effects on endemic species. If the problem is not effectively approached, extinction may occur. In an effort to control the issue at hand, restoration projects will be important in maintaining biodiversity. It is especially crucial to protect natural forests in areas with high biodiversity as a priority. It is also crucial to stop all current activity in converting forests for plantations with strictly enforced policies and the establishment of "environmental friendly corridors" in order to maintain connectivity, limiting the impact of fragmentation and isolation.

Conclusions

Tropical Rainforests are being cut down at a rate faster than can be recovered. There have already been serious impacts on climate change and biodiversity as a result of losing much of the world's forests already. Anthropogenic sources make it much harder to try and recover them, especially in consideration of the economies of the poor countries that live around the rainforests. However, it is not entirely hopeless, with stronger policies and better implementation of them it might be possible to recover much of these landscapes without too much damage to those who depend on the resources of tropical rainforests. Although more studies must be conducted in order to try and restore these environments, it is possible to save these beautiful areas.

References Cited

Bieber AGD, Silva PSD, Sendoya SF, Oliveira PS 2014 Assessing the Impact of Deforestation of the Atlantic Rainforest on Ant-Fruit Interactions: A Field Experiment Using Synthetic Fruits. PLoS ONE 9(2): e90369. doi:10.1371/journal.pone.0090369

Brando, P. M., Balch, J. K, Nepstad, C. N.,

Morton, D.C, *et al.* 2014 Abrupt Increases in Amazonian Tree Mortality Due To Drought–Fire Interactions PNAS 111 17, 6347–6352

Busch, J., Ferretti-Gallon, K., Engelmann, J., Wright, M., 2014, Reductions in emissions from deforestation from Indonesia's moratorium on new oil palm, timber, and logging concessions doi: 10.1073/pnas.1412514112

Carrara, E., Arroyo-Rodríguez, V., Vega-Rivera, J. H., Schondube, J. E.,(2014) Impact of landscape composition and configuration on forest specialist and generalist bird species in the fragmented Lacandona rainforest, Mexico. Biological Conservation 184, 117–126

Forrest, J. L., Mascia, M. B., Pailler, S., Abidin, S. Z., Araujo, M. D., Krithivasan, R. and Riveros, J. C. 2014, Tropical Deforestation and Carbon Emissions from Protected Area Downgrading, Downsizing, and Degazettement (PADDD). Conservation Letters. doi: 10.1111/conl.12144

Gonzalez, P., Kroll, B., Vargas, C., 2014, Tropical rainforest biodiversity and aboveground carbon changes and uncertainties in the Selva Central, Peru, Forest Ecology and Management, 312, 78–91

Pavlacky Jr., D.C., Possinghan, H. P., Goldizen, A. W. Integrating life history traits and forest structure to evaluate the vulnerability of rainforest birds along gradients of deforestation and fragmentation in eastern Australia. Biol. Conserv. 2014, http://dx.doi.org/10.1016/j.biocon.2014.10.020

Zhai, DL., Cannon, C., Dai, ZC., Zhang, CP., Xu, JC (2014) Deforestation and fragmentation of natural forests in the upper Changhua watershed, Hainan, China: implications for biodiversity conservation. DOI 10.1007/s10661-014-4137-3

Section III—Human Issues

The Social Cost of Carbon

Makari Krause

Climate change has already begun to have obvious and measureable effects on our planet. An increase in severity and frequency of storms and droughts coupled with rising sea levels and many other effects are imposing many costs on society. These costs are beginning to rise, and yet very little is being done to combat climate change. Fossil fuel companies are currently directly subsidized to the tune of $550 billion dollars worldwide. These companies are also subsidized in a more indirect and yet just as real way; they do not have to pay for the negative impacts that result when we burn the products they sell. This situation has led to a massive negative externality in which society will bear the brunt of the impacts generated from greenhouse gas emissions and the fossil fuel companies will get off scot-free. The amount that society will pay is commonly known as the social cost of carbon (or carbon equivalents when we take into account the other greenhouse gases) and is generally measured per ton of carbon emissions.

This is not an uncorrectable situation; negative externalities are commonly corrected through policies and society benefits as a result. An interesting question to ask, then, is why this situation persists if the majority of the population would stand to benefit from an internalization of this externality and only a few oil companies stand to lose. The answer to this question is that there is actually little per capita incentive for the majority of the population to push for such a policy. While the cost inflicted on society will no doubt be huge, this cost is widely dispersed across the population and no one man or woman stands to gain much from a policy limiting emissions. Those in opposition to the policy, however, have everything to lose. This results in the situation that we have today. Where a small but politically and financially powerful group of corporations with a lot to lose band together to block policies that would benefit society as a whole but would strip them of the subsidies they currently enjoy.

Taxes are one of the most powerful tools available to governments to internalize negative externalities. Currently the cost of fossil fuels simply reflects the direct costs to the firm, the costs associated with the use of those fossil fuels is not at all accounted for. This leads society to use them irrespective of the damage they are causing. Society, therefore, consumes more than the optimal level of these fossil fuels and in turn inflicts great damage on itself. A tax that raised the per unit cost of fossil fuels so that that cost equaled the sum of the cost to the firms and the cost to society would induce society to consume less fossil fuel and thereby diminish the impact to society. While the principle is straightforward, the enactment of such a policy is extremely difficult.

The first problem comes in determining the social cost of carbon. Estimates range from less than $10 per ton to hundreds or even thousands of dollars per ton. The reason for these differences is twofold. First, our physical models for the impact of an additional ton of carbon on the world are highly speculative. While it is undisputed that an additional ton of carbon will increase the greenhouse effect and raise global temperatures, it is unknown how much that ton will raise global temperatures. Even if that value were known, it is unknown how that temperature rise would translate to physical damage (to human infrastructure or to ecosystems), the values that we care about at the end of the day. Climate models are simply a rough approximation of reality and, especially when modeling far into the future, are highly uncertain. This reality makes it very difficult for policymakers to act on the science. How does one implement a carbon tax, even if everyone agrees that one needs to be implemented, when it is impossible to determine what its value should be?

Second, there is disagreement about the discount rate that should be used. Essentially the idea behind the discount rate is that a dollar today is worth more than a dollar next year. If you are given a dollar today you can invest that dollar, earn interest, and ideally have more than a dollar next year. This makes one dollar in the present more valuable than one dollar in the future. This principle means that we, as society, are unwilling to pay the full price today to avoid damages that may occur 100 years from now. We can instead invest that money with the hope that the return on that investment will aid in covering future damages when they occur. The discount rate is economists' way of including this principle into the social cost of carbon. However just as there was disagreement over the cost of damages, there is disagreement over the discount rate. Values range from 1% to 5% and can significantly alter the social cost of carbon.

Beyond the problem of establishing a value for the social cost of carbon, a carbon tax is a politically unfavorable move, even more so than other types of taxes. This is because the burden of the tax falls on current generations while the benefits of the tax are experienced by future generations. As a society focused mostly on improving our current personal welfare, many seem to believe that the benefits of the carbon tax do not outweigh the costs.

The next avenue for internalizing the externality is to simply eliminate our emissions from fossil fuel use. Obviously we cannot eliminate their use altogether but any decrease in emissions is a decrease in the negative externality. This decrease can be accomplished through a regulatory cap on total emissions. Again this is simple in principle but implementing the regulations becomes very challenging. Many high emitting sectors such as transportation and electricity generation are critical to the functioning of our society. These sectors must be permitted a certain level of emissions but who is to determine the level of emissions that are to be permitted. Furthermore, who is to determine the level of emissions granted to each firm? Emissions from power plants depend heavily on the type of fuel being burned and the technology used. Therefore one cannot simply allot each plant the same level of emissions. This becomes a regulatory nightmare.

Considering all of these challenges, it is clear why there is so much disagreement when it comes to the social cost of carbon. This chapter examines some of the research that has gone into determining the social cost of carbon and what can be done to internalize the externality, be that through a carbon tax that brings the real cost of carbon in line with the social cost of carbon or though a cap on emissions.

Risk Mitigation and the Social Cost of Carbon

In 2009 the Obama administration convened a working group to determine the social cost of carbon. To achieve this goal the group used three main models: Nordhaus' (2008) "Dynamic Integrated Climate Economy"(DICE) model; Hope's (2008) "Policy Analysis of the Greenhouse Effect" (PAGE) model; and Anthoff and Tol's (2010) "Climate Framework for Uncertainty, Negotiation, and Distribution" (FUND) model. The group decided to use discount rates of 2.5%, 3%, and 5% in each of the models. These discount rates were chosen based on a review of the literature.

Under the 5% discount rate the group determined that the social cost of carbon was $4.7 per ton in 2010. Under the 2.5% rate the cost of carbon was $35.1 per ton. Many critics argued that these values were based on overly optimistic models so the group subsequently used a fat tailed distribution on climate sensitivity and came up with $64.9 per ton at a discount rate of 3%. This number still falls far short of many other estimates.

Howarth *et al.* (2014) argue that appropriately accounting for risk mitigation might substantially increase the social cost of carbon above previous estimates. The authors use a previously developed model to show that greenhouse gas emissions mitigation is incredibly valuable up until the point at which catastrophic impacts are avoided. Once this level has been reached the model showed that there was low marginal benefit to any further reductions. These results suggest that the social cost of carbon therefore depends heavily on the stringency of emissions reduction policies. Under all scenarios emissions are completely mitigated by the year 2270. Without any mitigation measures market forces act slowly to accomplish this mitigation and atmospheric greenhouse gas levels peak at 740 ppm in 2100. With stringent mitigation, emissions begin to drop off immediately and atmospheric levels peak at only 460 ppm by mid century. Under the "no mitigation" scenario there is a 0.33% chance of climate catastrophe under the fat tailed assumptions about climate sensitivity. (Howarth *et al.* take climate catastrophe to mean that the standard of living is driven down to the subsistence level at some point in the next 400 years.) Under thin tailed assumptions this percentage drops to 0.025%. While these percentages are very low, Howarth *et al.* used a study that showed that peoples' risk aversion is high enough that these catastrophic events, no matter how unlikely, are enough to warrant stringent controls on emissions.

Howarth *et al.* conclude that with low risk aversion, the social cost of carbon is $10 in 2010. Using the observed risk aversion in financial markets, however, the social cost of carbon in 2010 is $25,700 for fat tailed uncertainty and $1,690 for thin tailed uncertainty. In both of these scenarios the cost of carbon drops off quickly once the risk of catastrophic climate events are mitigated. Howarth *et al.* suggest that moving from the no abatement situation to one in which there is stringent abatement could produce benefits equal to a four fold increase in per-capita consumption. This is because abatement measures can decrease the probability of catastrophic climate events to essentially zero and are really the only way to deal with these catastrophic events. Traditional forms of risk mitigation will not work because these climate events will affect the economy as a whole. Once these catastrophic events have been avoided, Howarth *et al.* found that there is low marginal benefit in additional mitigation.

The Economic Impact of Extreme Sea-level Rise

Sea level rise is one of the most concerning facets of climate change, Pycroft *et al.* (2014), examine its effects on the social cost of carbon. Rapid ice sheet melting or collapse which would cause rapid, significant sea level rise is hard to incorporate into climate models because it is difficult to gauge the likelihood of such events. This difficulty stems from the fact that the underlying processes are, themselves, hard to model. Pycoroft *et al.* use an integrated assessment model to examine the impacts and costs of large-scale damage associated with sea level rise. In their model they adjust the physical aspects that contribute to sea level rise and the economic consequences of those aspects. Their model shows that incorporating extreme sea level rise significantly increases the social cost of carbon.

Pycroft *et al.* begin with a review of economic and scientific literature surrounding sea level rise. There is continuing uncertainty about the extent of temperature rise and even more uncertainty about how changing temperatures will affect sea levels. Through looking at the literature Pycroft *et al.* determine that sea level rise of more than 1m would require a significant increase in ice flow dynamic and that sea level rise over 2 m is highly unlikely. For this reason their 90% confidence interval is between 1 and 2 m and sea level rise parameters have been adjusted accordingly for global temperatures. (The standard model uses values of 0.4 and 1 m)

The integrated assessment model used in this study, PAGE09, separates impacts induced by climate change into four categories: economic, non-economic, sea-level rise, and discontinuity. Economic and non-economic damages depend on temperature, sea-level rise damages are explicitly modeled through sea level rise, and discontinuity damages capture everything that doesn't fall into the other categories, such as catastrophic events like ice sheet collapse. Pycroft *et al.* calculate the social cost of carbon by first running the model as normal and then running it with a slight decrease in the amount of CO_2 in the first year. They then can calculate the marginal impact per ton of carbon dioxide and from that the cost.

Pycroft *et al.* replace discontinuity damages in the original PAGE09 model with tails, allowing for the possibility that economic damage estimates are much higher than the central estimates. Under the original PAGE09 model sea level rises by an average of 0.64 m by 2100 and 1.61 m by 2200. When tails are added to the model, sea level rise in 2200 increases by 8% (thin tail), 9% (intermediate tail), and 12% (fat tail). Under the revised sea level rise parameters mean sea level rise increases from 0.64 m to 1.43 m in 2100 and from 1.61 to 3.98 m in 2200. The revised parameters clearly increase the response of sea level to temperature. Translated into mean dollar values for the cost of carbon, the thin tailed parameters yield $135/ton, the normal tailed yield $147/ton, and the fat tailed yield $218/ton. When revised sea-level parameters are introduced these values change to $149/ton, 161$/ton, and $218/ton respectively.

The revised sea level parameters add between 10 and $14/ton of CO_2 to the mean values for the social cost of carbon dioxide. This estimate increases considerably when the revised sea level parameters are combined with the fat tailed distribution. In this case the 95th percentile estimate increases from $54/ton to $893/ton. This shows that the combination of revised parameters and fat tails gives a lot more weight to damages caused by rare, extreme climate events. More research needs to be done to determine the accuracy of the different tail sizes and determine which variant of the model will most reflect reality.

Carbon Price Analysis Using Empirical Mode Decomposition

Zhu *et al.* (2014) aim to enhance the science on properly setting and forecasting carbon prices. To do so they examine the European Union Emissions Trading Scheme (EU ETS) through empirical mode decomposition to better understand the formation mechanism of carbon prices.

The EU ETS is the largest carbon market in the world, covering 12,000 installations and 25 countries. Over the past few years there have been a number of studies analyzing carbon prices in the EU ETS, that have generally fallen into one of two categories; structured models and data-driven models. Structured models analyze carbon price movement through the perspective of supply and demand and can help with understanding the generation of carbon prices. However, because of the unstable nature of the market, these models have been difficult to implement. Data-driven models, such as linear regressions, work well for short-term forecasting but fail to explain the driving forces behind carbon price changes.

To avoid these limitations, Zhu *et al.* use Empirical Mode Decomposition (EMD) in their analysis. EMD is a data analysis approach for nonlinear and non-stationary time series that can decompose carbon prices into one residue function and several independent intrinsic mode functions (IMF). In this analysis Zhu *et al.* produce eight IMFs. Each of these is based on local characteristic scales and can be analyzed to better understand the characteristics of the carbon price and the underlying factors influencing that carbon price. After decomposition, the resulting IMFs and residue are reconstructed into a high frequency component, a low frequency component and a trend component. Economically speaking, the high frequency component is interpreted as short term fluctuations, the low frequency component is interpreted as significant trend breaks, and the trend is interpreted as the long term economic trend.

Analysis of the trend line shows the evolution of the carbon price in the long term. The low frequency component (significant trend breaks) can explain sharp ups and downs in the carbon price, such as jumps associated with changes in government policies. Finally the short term, high frequency component mostly describes normal market operations that drive market fluctuations. These three components of price can then be forecasted into the future and combined to produce future projections of carbon prices.

Political Economy Constraints on Carbon Pricing Policies

While the scientific community and much of the public accept the reality of climate change, little is being done to curb our emissions. There are a number of ways to get emissions in check. You can have command and control regulations, you can use economic means such as a cap and trade system for GHG emissions or a tax on GHG emissions, and you can also provide subsidies or set production quotas for low-carbon energy sources. A lot of research has been done on these different alternatives and unsurprisingly the studies usually find that carbon-pricing policies are the most efficient. In their paper Jenkins *et al.* (2014) refer to this method of pricing as the "first-best" response or the most cost effective way to accomplish emission mitigation. While economists argue that carbon pricing should be used alone in order to internalize the negative externality, this is rarely the case in the real world. Often pricing policies are combined with many other instruments such as energy efficiency incentives and regulations to meet goals. While this mix of policy instruments may not provide the theoretically most efficient solution, in the real world this theoretical solution may not be attainable and research has shown that a mix of policy instruments may actually provide a better result than using only carbon pricing instruments.

Much of the research showing this has looked at market failures and institutional limitations but there are other political economy constraints that are also a major factor when looking at the efficacy of a carbon pricing system.

One of the most significant constraints is that there are several sectors of the economy that would suffer heavily from the imposition of a carbon price and these sectors would mount significant opposition to the policy. Sectors such as steel, chemicals, fossil energy and mining have fixed assets that depend heavily on maintaining particular market and regulatory conditions. As a result, firms in these sectors face strong incentives to overcome collective action hurdles and work together to influence the regulatory process. With the costs focused largely on a small number of sectors and the benefits widely diffused throughout the population, these small groups are often very capable of shaping policy and blocking mitigation efforts.

Additionally carbon pricing will lead to loss in private welfare and a transfer of welfare to the government through tax revenues and away from consumers and producers. The shift in market equilibrium caused by the new carbon price would also impose costs even if welfare is eventually maximized.

Another major constraint on carbon pricing is that the costs of mitigation are felt directly by private consumers and citizens in the near term while the benefits are primarily felt by future generations and will be dispersed across the entire world. Costs of dealing with climate change will, in fact, be most heavily felt by citizens of the United States while citizens of poorer nations will feel the majority of the benefits. This results in a very low willingness to pay for mitigation. Polling has shown that consumer willingness to pay falls between $80 and $200 per household per year which translates to a carbon price of $2-$8 per ton. At this level, a viable carbon tax would far outstrip consumers' willingness to pay for mitigation even if the revenues from the tax were used to offset other federal taxes such as income tax. Carbon prices in associated literature fall between $12 and $150. This unwillingness to pay leads to a situation under the carbon price policy scenario that isn't environmentally efficacious or economically efficient.

In the face of the abovementioned constraints on climate policy, Jenkins *et al.* believe that second-best policies can achieve superior economic efficiency, environmental efficacy, and political feasibility than carbon pricing policies alone when applied in the real world. Jenkins *et al.* conclude their paper with a list of policy implications that should be taken into consideration when designing a carbon mitigation strategy:

1. Policies should be compared to an economically optimal carbon pricing instrument to evaluate their performance.
2. The choice of the policy mechanism itself can affect consumer willingness to pay.
3. Careful attention should be taken to neutralize opposition from energy-intensive manufacturers.
4. Mitigation strategies should aim to link long-term avoided climate damages with near-term benefits.
5. It should be kept in mind that economic and political constraints on the optimal climate policy are not static.
6. Constraints on carbon pricing instruments makes the creative use of resulting revenues critical to maximizing the economic efficiency and environmental efficacy of these instruments.

Tax Policy Issues in Designing a Carbon Tax

Carbon taxes have long been thought of as the most efficient and successful way to decrease GHG emissions and thereby curb climate change. Marron and Toder (2014) examine some of the challenges associated with this approach to carbon mitigation, namely setting the tax rate, collecting the tax, and using the revenue. In order to internalize the GHG emissions externality one must tax those emissions at a rate that brings the social cost in line with the private cost. This is referred to as the social cost of carbon and is the price of carbon that would maximize social welfare. Theoretically this approach seems ideal but there are many difficulties involved with determining the social cost of carbon. Determining the true economic effects of GHG emissions is quite difficult and requires complex modeling. These models operate on a set of assumptions that are controversial in many cases. Leading to a wide range of estimates for the social cost of carbon with a mean of $196/ton and a standard deviation of $322/ton. Another important question to ask when calculating cost is whether that cost will be evaluated on a global or national scale. The costs of climate change and the benefits of mitigation are global but often US policymakers exclude global considerations.

Actually collecting the carbon tax once the rate has been decided brings its own set of challenges. Ideally one would monitor all emissions and tax them at a uniform rate. This is impossible because the cost of monitoring millions of emissions sources is prohibitively expensive. Alternatively one could tax large sources of emissions such as the power sector but this would fail to capture many other sources. The most plausible way to collect the GHG tax is to instead tax fossil fuels directly. Fossil fuels account for 90% of carbon dioxide emissions in the US and their carbon content can be easily measured and translated into emissions after combustion. These fuels can be taxed at one point along the supply chain to increase the ease of monitoring and decrease the number of taxpayers. They can also be taxed at the retail level but this doesn't encourage increased efficiency in the refining and production processes.

For the tax to be comprehensive it would have to focus on other carbon-intensive industries such as chemicals and steel in addition to fossil fuels and on GHGs other than carbon dioxide. Taxes would also need to be rebated in the case that the fossil fuel didn't result in emissions. This would be necessary if carbon capture and sequestration became widespread in the power production industry.

The last question that further complicates the adoption of a carbon tax is what to do with the revenues. Because lower income households spend relatively more on carbon-intensive products than do upper income households, a carbon tax would be regressive and most of the burden would fall on lower income households but carbon tax revenues could be used to provide tax relief and offset the regressive nature of the tax. Additionally carbon tax revenues could go towards research and development of clean, renewable energy sources. Some of the revenues could also be used to assist communities that rely heavily on carbon-intensive industries such as the families of coal miners. Obviously the revenue could also go towards reducing the federal deficit.

A tax on carbon, while theoretically optimal, will obviously be very hard to implement and there are many considerations that need to be addressed.

Regulating Greenhouse Gas Pollution from Existing Power Plants

One of the most ambitious components of President Obama's Climate Action Plan is the EPA's Clean Power Plan (CPP). The CPP combines regulations on new and existing power plants and will drastically reduce power plant emissions once implemented. Authority for the CPP is granted by Section 111(d) of the Clean Air Act and requires that the EPA issue a set of emissions guidelines. Once they have done this the states then need to come up with their own way of meeting those emissions guidelines but have discretion in deciding what instruments to use as long as the resulting abatement is either the same or superior to that mandated by the EPA.

There are generally two areas of contention in this process. First are concerns about EPA's establishment of the emissions guidelines. Different parties have different opinions about what form the guidelines should take, whether they should be centered around mass-based caps or rate-based performance standards. There is also disagreement about how EPA should define the best system of emissions reduction. Should EPA consider carbon trading and demand side efficiency as a way to meet the targets or require that the targets be met by technological improvements within the plants themselves? And of course there remains the ever–present question of how much abatement the EPA should be aiming for with the program.

The second area of contention is about the amount of flexibility the states should be afforded when designing plans to meet the EPA requirements. Should they be able to undertake carbon trading or carbon offsetting projects or be limited to technological advancements and demand side efficiency improvements?

Lienki *et al.* (2014) include a chart summarizing the positions taken in 30 reports on the questions and concerns raised above. The reports were produced by environmental groups, industry, and government agencies. Three main broad themes become prevalent when looking at these reports:

Embrace Broad, Market Based Compliance Mechanisms

Most stakeholders agree that EPA should afford states a lot of discretion and flexibility and should allow the use of emissions trading and averaging, demand side efficiency, and increases in renewable energy generation to meet abatement goals. Most stakeholders do not, however, believe that carbon offsets should be allowed.

Encourage Mass Based Standards

The majority of proposals also recommend that EPA allow states to use mass-based emissions caps rather than rate-based performance standards. This does not necessarily mean that the EPA must use mass-based caps when determining the emissions guidelines, however. The reason for this is that many economic studies have shown mass-based cap-and-trade systems to be more efficient than rate-based trading systems. Mass-based trading programs put an effective price on every ton of carbon emitted. If this price were appropriately calibrated to equal the social cost of carbon then all cost-effective emissions reductions would take place. A rate-based system, however, only puts a price on emissions above the relevant performance standard. Mass-based systems also allow demand-side energy efficiency improvements to be easily taken into account because they decrease the total level of emissions. These improvements are much harder to account for in a rate-based system because efficiency improvements do not decrease the rate at which generators produce carbon. This same argument holds for generation from non-carbon emitting sources.

The authors conclude with a recommendation that EPA conduct a full-cost benefit analysis to identify the social cost of carbon and thereby the optimal level of emissions reduction to call for in the CPP.

Can Carbon Taxes be Progressive?

Dissou *et al.* (2014) provide a new assessment of the incidence of carbon taxes and their impacts on inequality. Past studies on the impacts of carbon taxes have generally focused on the effects to commodity prices. This channel usually leads to the finding that carbon taxes are regressive because a carbon tax increases the price of energy-intensive goods and these goods make up the largest portion of the total goods bought by poorer households. Using only this one channel, however, overlooks a critical aspect of the tax; factor income, income generated from selling factors of production such as labor, is also affected by carbon taxes. Dissou *et al.* combine both the commodity price channel and the factor price channel to get a more comprehensive picture of the effects of a carbon tax on inequality and find that carbon taxes are not as regressive as previously thought.

The rich derive much of their income from capital, the poor on the other hand derive much of their income from labor. Past studies have shown that pollution control policies can harm capital income more than labor income because polluting industries are generally more capital-intensive than other industries. The outcome of a pollution control policy could, therefore, be progressive as it harms the rich more than the poor.

Dissou *et al.* decompose welfare metrics into three different components; initial total expenditures, contribution from commodity price changes, and contributions from changes in factor prices (i.e. wages). They then create a model and run several simulations with different carbon tax values: $15, $30, $50, $100, $150 per ton. The model produces the impact on commodity and factor prices, which are then used to measure the impact on inequality.

The first direct impact of the carbon tax is an increase in energy prices. This leads to decreased demand and lower carbon emissions. Increasing energy prices also leads to higher costs to firms and therefore higher prices in all industries. The magnitude of this price increase depends on the industry but it is this change in the price of goods that makes the tax more regressive. As the marginal cost increases, the returns to labor and capital decrease. This decrease is larger for capital and because capital is used more heavily in energy-intensive industries, output will fall in these industries more than in others. This will lead to a decrease in the demand for capital relative to labor and an increase in price for labor relative to capital. This factor price change makes the tax progressive

Results show that there are two clearly opposing effects on inequality the stem from a carbon tax. On the one hand the changes in factor prices lead to a more progressive tax in which wealthy households are impacted more heavily than poor households. On the other hand changes in commodity prices lead to a more regressive tax. The net impact depends on the strength of each of these opposing effects and is difficult to infer.

In their simulated economy, Dissou *et al.* did, however, observe an interesting phenomenon. At low levels, carbon taxes were found to be progressive but as the taxes increased they were found to be increasingly regressive. The distributional impact of the carbon tax might therefore be dependent on both the level of the tax and on the structure of the economy in question.

A New Strategic Plan for a Carbon Tax

Stram (2014) proposes a new greenhouse gas policy that is much more politically acceptable than current alternatives. The policy centers on a small carbon tax, the revenues of which go towards research and development of clean energy sources and mitigation. It also offers incentives for other nations around the globe to participate.

According to Stram, cap and trade programs for greenhouse gases have generally failed. These types of programs have been successful in the past with sulfur oxide and nitrogen oxide but not with carbon. Additionally these programs took place within the US and therefore might not be comparable with the global solution that is needed for greenhouse gasses. There will not be an organization like the US EPA overseeing a global GHG policy and implementation and compliance enforcement will be much more difficult. In the face of these challenges, Stram proposes a new, more moderate strategy that may be acceptable to skeptical politicians. The strategy hinges on the adoption of new, cheap, low- or no-carbon energy technologies to provide our energy needs. The path to achieving this goal is through a small carbon tax, which, if levied across some of the largest emitters in the world would provide ample funds for research and development. Therefore, even if the tax itself does not decrease emissions to the desired level it would provide a channel through which that decrease can be achieved by market forces once new energy generation technology is developed.

The policy should first be set up with a very small tax. The goal here is to avoid major opposition to the policy while the general structure is being implemented. Later the tax can be increased as needed. Even a mere $2 per ton tax in the US alone would generate $12 billion. If other countries join, the tax revenues would be significant. The incentives for other countries to participate in the program comes from Stram's underlying assumption that cheaper energy sources are available if enough money is put into researching and developing them. Countries contributing to this program would have a share in the benefits that eventually come from the cheaper energy source that is developed.

While many will criticize this approach as not going far enough, Stram argues that current alternatives are not working and a small carbon tax administered in the way described above is better than the status quo and, if his assumptions hold, will solve the problem nonetheless.

Conclusions

While the social cost of carbon is still very much undetermined, the literature on the topic continues to grow and with each passing year more research is contributed. As modeling advances, the costs associated with climate change will become clearer and policies combating climate change will become more politically feasible. The main question is whether the science can advance fast enough so that something can be done before the effects of climate change become vast and irreversible. Emissions continue to grow and with rapid development in many parts of the globe, this trend is only increasing. Scientists and policymakers across the globe need to come together and formulate a plan that is both politically acceptable and robust enough to halt climate change.

References Cited

Dissou, Y., & Siddiqui, M. S., 2014. Can carbon taxes be progressive? Energy Economics 42, 88-100.

Howarth, R. B., Gerst, M. D., & Borsuk, M. E., 2014. Risk mitigation and the social cost of carbon. Global Environmental Change 24, 123-131.

Jenkins, J. D., 2014. Political economy constraints on carbon pricing policies: What are the implications for economic efficiency, environmental efficacy, and climate policy design? Energy Policy 69, 467-477.

Lienke, J., Schwartz, J. A., & No, P. B. (2014). Regulating Greenhouse Gas Pollution from Existing Power Plants.

Marron, D. B., & Toder, E. T., 2014. Tax policy issues in designing a carbon tax. The American Economic Review 104, 563-568.

Pycroft, J., Vergano, L., & Hope, C., 2014. The economic impact of extreme sea-level rise: Ice sheet vulnerability and the social cost of carbon dioxide. Global environmental change 24, 99-107.

Stram, B. N., 2014. A new strategic plan for a carbon tax. Energy Policy 73, 519-523.

Zhu, B., Wang, P., Chevallier, J., Wei, Y., 2014. Carbon price analysis using empirical mode decomposition. Computational Economics 45, 195-206.

Climate Change and Urban Development

Dan McCabe

Cities and their surrounding suburbs, which house most of the world's population, present a unique challenge to officials, scientists, and engineers aiming to enhance global sustainability. As the centers of global population and commerce, urban areas face two distinct sustainability challenges. First, cities are directly and indirectly responsible for most of the world's resource usage and emissions of greenhouse gases and other pollutants. Second, highly developed areas must deal with certain local environmental issues unique to the built environment. The urban heat island effect, for example, can increase the daytime temperatures of cities by several degrees Fahrenheit compared to natural areas with the same climate (Rotem-Mindali *et al.*, 2015). This well-documented phenomenon occurs because built surfaces reflect and absorb sunlight differently than vegetated spaces do, resulting in a net warming effect. A number of other distinct urban environmental concerns, such as flood management (Barau *et al.*, 2015), need to be taken into account as well. To make matters worse, global and local sustainability concerns in cities tend to add together to increase their severity. The temperature increase due to the urban heat island, for example, places additional stress on the cooling systems of local buildings, substantially increasing their energy usage and greenhouse gas emissions.

The studies addressed in the following chapter aim either to assess the state of different urban environmental problems or to evaluate different strategies designed to counteract them. Included are factors that drive consumption and carbon footprint (Creutzig *et al.* 2015; Jones & Kammen 2014), vulnerability to natural disasters (Barau *et al.* 2015), urban heat island quantification (Rotem-Mindali *et al.* 2015; Baró *et al.* 2014), evaluation of climate change mitigation methods (Rotem-Mindali *et al.* 2015; Baró *et al.* 2014; Georgescu *et al.* 2014), and other analyses. Much of the research aims to determine what factors most strongly influence various sustainability criteria; linear regression and correlation analyses are often used to investigate the relative weight of many possible variables. In some cases, these studies are done in order to unveil the most important factors to be targeted by mitigation and adaptation strategies; other times, the impacts of different proposed solutions to urban environmental problems are evaluated.

Across the planet, urban planners, policymakers, engineers and scientists are working to develop solutions to worsening environmental problems due to urban development. It is currently highly difficult to determine the appropriate mitigation and adaptation methods because it is still unclear just what factors shape sustainability in urban areas. Continued research hopes to provide more insight into what makes cities sustainable. These efforts are becoming increasingly important in light of rapid population growth and urban expansion. Intelligently planning the development of new cities and managing the growth of established urban regions can greatly reduce the potential negative impacts of greater population, a daunting challenge that needs to be addressed as soon as possible.

Vegetation Level and Surface Temperature in Tel Aviv, Israel

One key objective of sustainable urban planning is to limit the urban heat island (UHI) effect, the increased local temperature in highly built areas due to differences from the natural environment in the absorption and reflection of solar energy at the surface. Previous research has displayed the value of large urban parks in controlling temperature in cities, but less is known about the effect of smaller green spaces. In order to investigate how vegetation and construction levels impact UHI severity, Rotem-Mindali *et al.* (2015) used ten years of remotely sensed data from two NASA satellites to analyze the relationship between different land uses and land surface temperature (LST) in Tel Aviv, Israel. The authors compiled information on local LST and Normalized Difference Vegetation Index (NDVI), a measure of vegetation cover, and used it to search for a correlation between land use type and mean surface temperature for summer nights. In their analysis, they found an enormous difference of 13°C in mean temperature among different locations in Tel Aviv. There was a strong correlation between land use type and LST, with the most vegetated regions experiencing much lower average temperatures than highly built regions.

Tel Aviv is Israel's largest metropolitan area and contains a diverse array of land use types. For the purposes of their analysis, the authors separated these into four categories: "green" residential areas characterized by prevalent vegetation in private gardens, other residential spaces, industrial regions, and small- to medium-sized public parks. Five plots of each type, all close to the city center, were selected for analysis. In addition to the results mentioned earlier, one of the study's most significant and unprecedented findings was that residential areas with substantial vegetation are in some ways more effective at offsetting UHI than small parks. Green residential districts exhibited mean LST 0.5°C less than these parks, but more significantly, while the thermal benefits of the parks diminish rapidly in their immediate surroundings, the UHI offset due to green residential areas is more spatially constant. The authors acknowledge that this result is not simply a testament to the superiority of private gardens for heat mitigation, because parks tend to be surrounded by parking lots and larger streets that generate a large thermal load. However, some of the advantage of residential green areas may be attributable to a more efficient allocation of plant life. The dense planting of trees and shrubs in residential areas and more uniform distribution of vegetation may enhance their mitigation capabilities.

The authors' finding that residential private gardens are highly effective at UHI mitigation is encouraging for the future of urban planning for climate management. In a dense and established urban fabric, it can be difficult to develop or incorporate large new green spaces that promote a more moderate local climate. This research, however, finds that the increased prevalence of urban gardens in residential areas can provide similarly significant benefits. Thus, the development of more green neighborhoods in cities can generate "cooling islands" to help promote more moderate urban temperatures that reduce energy demand and improve quality of life.

Models Reveal Climatic Impacts of Urban Expansion

In order to quantify and analyze the impacts of urbanization on local and regional temperature and hydroclimate, Georgescu *et al.* (2014) modeled the impacts of urban expansion in the contiguous United States in a variety of scenarios. The authors considered a range of different predicted population levels in the United States for the year 2100. Using advanced atmospheric models, they found that if no urban climate change mitigation measures were put into place by then, summertime urban-induced warming of 1–3°C can be expected in cities, with exact values varying by location. These increased temperatures are due solely to the effects of the built environment, as simulations were run using climate data from 2001–2008 without any assumptions about future warming due to increased greenhouse gas emissions.

The model also evaluated the impacts of certain urban climate adaptation strategies—specifically, the implementation of vegetated green roofs, reflective cool roofs, and a hybridization of the two roof styles. The results suggested that widespread implementation of these strategies could entirely offset the warming caused by urban development. While this result is promising, it came with some caveats. For example, the authors found that widespread adoption of cool roofs counteracts summertime warming but exacerbates winter cooling, leading to increased heating demand and energy usage in winter. The alternative roofing styles also exhibited substantial effects on hydroclimate, as cool roofs showed a tendency to decrease precipitation and green roofs to increase it. Interestingly, these impacts appeared to extend beyond local areas and have noticeable effects on regional hydroclimate, which need to be addressed in any future plans for urban climate change adaptation.

The authors' research emphasizes a need for comprehensive responses to climate change. While the model revealed many universal warming trends for urban areas, individual cities suffered unique consequences. However, the need to address the effects of urban warming is clear. In the case of maximum predicted greenhouse gas emissions by 2100, regional contribution to urban warming is expected to range from 15–27% (up to 50% locally), and is even higher in the case of lower emissions. The results of this research demonstrate that local and regional climate change due to urbanization is likely to be a problem regardless of global trends, and the policy needed to combat it should comprehensively address climatic, hydrological, and socioeconomic concerns unique to the regions it covers.

The Benefits and Limitations of Barcelona's Green Space

One aspect of urban ecology that is often overlooked in development is the biological benefit of vegetation in cities. In order to quantify the environmental impact of urban plants, Baró *et al.* (2014) analyzed the effect of green spaces on air quality and carbon sequestration in the city of Barcelona, Spain. The authors randomly selected nearly 600 small plots of land within the city limits and collected field data on the plant life and pollutant levels in each. This information, along with meteorological data, was then processed using i-Tree Eco software, which quantified the biological and economic effects of vegetation on both air quality and climate change. In this software model, green space is treated as providing two kinds of ecosystem benefits—defined as air purification and global climate regulation—as well as one harmful consequence, the emission of biogenic volatile organic compounds (BVOCs). For this case study, the model focused only on the levels of particulate matter (PM_{10}) and NO_2 and no other pollutants that harm air quality, because Barcelona has recently had exceedingly high concentrations of these two pollutants.

The i-Tree Eco model found that the impact of green spaces in Barcelona was substantial both for air quality improvement and for carbon sequestration, but small relative to total emissions. Air pollutant removal totaled hundreds of tons, but this figure amounted to only 0.52% of Barcelona's total NO_2 emissions and 22.31% of PM_{10} emissions—but less than 3% of urban PM_{10} levels when including pollution that did not originate in the city. Likewise, climate change mitigation due to carbon sequestration in plants was substantial, estimated at nearly 20,000 tons of CO_2 per year, but this figure represents only 0.47% of Barcelona's annual citywide greenhouse gas emissions.

The authors' results display the benefits that urban vegetation has on air quality and carbon emissions mitigation, but also demonstrate its limitations. Although urban forests appear to have a meager impact on airborne pollutants and greenhouse gas emissions in Barcelona, they can still be a valuable tool for urban planning. Green spaces in cities may not be able to offset air pollution and climate change on their own, but they can provide other services such as recreation opportunities and local climate regulation while simultaneously capturing harmful pollutants and carbon emissions, which promotes human health and can save millions of dollars. Mindful management and expansion of urban green space therefore appears to be a viable strategy for fighting air pollution and climate change, but a truly successful urban sustainability plan must integrate green space preservation and development into a multifaceted plan to enhance sustainability.

The Underappreciated Contributions of "Green" Urban Brownfields

Sustainable urban planning aims to ensure that urban development patterns do as little ecological harm as possible, but new research suggests that conventional planning approaches neglect a significant contributor to urban environmental health. Mathey *et al.* (2015) studied the effects of urban brownfields, formerly developed sites that have been abandoned and remain underused, and found that vegetation-rich "green" brownfields provide a variety of valuable ecological services to their areas. These benefits depend on the type and amount of vegetation at a particular site, its location, and human intervention, but green urban brownfields overall show a large potential to aid the goals of sustainable development. Specifically, brownfields provide habitats that support enhanced biological diversity, aid microclimate by cooling the unnaturally warm urban environment, and provide local residents with recreational opportunities. These findings were obtained via a literature review, climate modeling, and a survey of local residents. The authors concluded that green urban brownfields should receive more credit for their ecosystem services and the most effective way to reap their benefits is to leave them mostly wild, while possibly converting some areas to recreational spaces.

Green urban brownfields have a particularly large potential to promote sustainable urbanism because of their prevalence; they are extremely common in cities that have transitioned from industrial economies or undergone population decreases. Over time, abandoned sites are colonized by plant life that expands and attracts an array of animal species, which can sometimes be threatened or endangered. Conventionally, brownfields are seen as potential locations for urban redevelopment, but the authors considered the implications of either converting them to more conventional green spaces or leaving them essentially wild. A review of case studies of the ecology of green urban brownfields in Germany was used to assess the habitat services provided by brownfields of different vegetation levels. The authors found that species diversity tended to increase along with the age of the site and the structural complexity of its plant life.

In an effort to quantify the microclimate services of brownfields, the authors used climate modelling software to evaluate surface temperature in the area surrounding these sites. Green spaces tend to provide a cooler environment than built areas, and vegetated brownfields are apparently no exception. While densely wooded areas may cause cooling of up to 2.1 K, the effects of brownfields are substantial but slightly less pronounced, ranging from 1.4 K to 1.7 K depending on vegetation level.

The final part of this study consisted of a survey given to residents of Dresden, Germany, on their use and perception of brownfields in the area. Respondents shared that they tend to use brownfields for walking, meeting places, experiencing nature, and more. They also expressed, however, a strong preference for traditional well-maintained urban parks as a location for recreational activities. Thus, while brownfields do appear to provide some recreational services to residents, the benefits are limited.

The results of this study were largely qualitative, but the authors were still able to draw some valuable conclusions. To examine the implications of different planning policies, they applied the knowledge obtained to two different scenarios in a brownfield-rich area in Dresden. In the first scenario, large-scale rebuilding took place; in the second, green brownfields were left relatively wild or converted to conventional maintained green spaces. The second scenario resulted in cooler temperatures, greater ecological diversity, and more recreational opportunities for residents. While the authors did not provide concrete advice for urban planning strategies regarding brownfields, their research illustrates the undervaluation of these green spaces in conventional urban planning and reveals the benefits they provide. More research is needed to investigate the ideal planning strategies regarding brownfields, but this paper lays the groundwork for a more comprehensive planning approach that takes full advantage of these unique and valuable spaces. Because brownfields are often the only green spaces available in urban areas, properly exploiting their ecosystem services can have tremendous benefits for local ecology.

The Anthropogenic Roots of Increased Flooding in Kano, Nigeria

Intelligent planning for urban development requires an understanding of how different development paths can impact sustainability. In order to better understand what aspects of cities impact sustainability, Barau *et al.* (2015) investigated historical trends in the environmental resilience of Kano, Nigeria. Kano, northern Nigeria's largest city with a population of over 2 million, has been a commercial center since the 10th century and has experienced extreme morphological changes in the centuries since then. Recently, the city has been subject to an increasing number of catastrophic flooding events that have caused deaths, exacerbated the spread of infectious diseases, and forced the relocation of hundreds of thousands of residents. As the frequency of extreme weather threatens to increase due to global climate change, Kano's ability to respond to flooding is of great concern. Barau *et al.* therefore sought to determine how the city's evolution has made it especially prone to severe floods.

In order to determine the relationship between Kano's development and environmental problems, the authors examined an assortment of historical images and texts to gain understanding about the city's structure. Their research was guided by the coupled human and natural systems (CHANS) framework, which is centered on the idea that anthropogenic and environmental variables combine to influence urban sustainability in a nonlinear way. From their analysis of both modern and historical sources, the researchers found that Kano has undergone extremely rapid spatial change between 1826, the date of the earliest informal map of the city available, and the present day. Specifically, natural building and roofing materials (primarily mud and clay) gave way to structures of impervious cement and metal. At the same time, streets came to take over more of the city area and became paved, ponds in the city decreased in number and size, and once-abundant green space was almost entirely eliminated. Overall, it was obvious that there was a drastic decrease in the prevalence of structural and ecological features that previously helped the city tolerate flood conditions. Western colonization brought along technology that altered the development of Kano's highly natural landscape, and eventually the relatively natural state of the city gave way to urban planning and engineering methods that removed most of Kano's open space and water. This transformation led to decreased rainfall absorption and greater runoff, turning rain events that once were manageable into more severe floods.

In addition to presenting insight into the forces that have caused environmental problems in Kano, the research demonstrates the importance of interdisciplinary approaches to assessing the environmental repercussions of urban development. Land use in cities is driven by policy, technology, history, culture, and more, all of which interact with one another nonlinearly, so it is necessary to holistically assess human impacts on the natural environment. In the Global South in particular, where cities' ancient structures impact their modern-day layouts, a thorough consideration of what shapes development is crucial. Understanding this relationship will be vital to global sustainability as the populations of cities across the planet continue to grow.

Intelligent Planning Can Limit Increases in Energy Demand

Urban areas account for the majority of global energy consumption and greenhouse gas emissions, which is of growing concern as their populations are projected to double within the next 35 years. In order to inform urban planning efforts to reduce greenhouse gas emissions, Creutzig *et al.* (2015) studied how a wide array of variables influence the energy consumption of cities across the globe. The authors considered detailed data provided by the World Bank (WB), the Global Energy Assessment (GEA), and the International Association of Public Transport (UITP) for 274 different-sized cities from 60 different countries. A correlation analysis was performed to determine how significant an impact each variable—such as gasoline price, population density, and gross domestic product (GDP)—appeared to have on citywide energy consumption. The dependent variable for this analysis depended on the data set from which information was obtained: per capita energy use for the GEA data, per capita transportation energy use for the UITP data, and per capita greenhouse gas emissions for the WB data. A standard linear regression model was used to determine the significance of each independent variable.

The results of the analysis depend on the data set and corresponding dependent variable considered. GHG emissions and transport energy consumption are most strongly correlated with gasoline prices and population density. Total energy consumption, on the other hand, is most affected by urban economic activity and climate, measured in heating degree days. The influence of each possible variable, however, differs across the range of cities studied.

An understanding of the important factors that contribute to urban energy demand allowed the researchers to assess the potential to reduce consumption. The authors found that strategic urban planning and the ideal implementation of fuel taxes could reduce projected urban energy consumption in 2050 from about 730 EJ in the business-as-usual scenario to 540 EJ, with most of the difference coming in the developing regions of Asia, Africa, and the Middle East.

While the analysis in this paper does not perfectly explain the energy usage of vastly different urban areas and projections for the future always carry substantial uncertainties, it makes some highly valuable contributions to understanding how cities can be more sustainable. Specifically, the energy demand of cities in the coming decades can be reduced by promoting high population density and, when possible, higher fuel prices. Unique planning strategies should be customized for different areas according to their needs; for example, fuel taxes are more effective in relatively affluent locations. While there does not appear to be a way to prevent consumption from rising severely as global population grows, this research offers pragmatic, useful data to encourage cities to develop with as little impact as possible.

Large Suburban Carbon Footprints Negate GHG Benefits of Urban Areas

Jones and Kammen (2014) performed a remarkably thorough analysis of the average household carbon footprint (HCF) for nearly every US zip code and examined how dozens of different variables affect greenhouse gas (GHG) emissions. The authors' analysis used detailed data from the nationwide Residential Energy Consumption

Survey, the National Household Travel Survey, and other sources. Their model used these surveys to estimate local emissions due to components such as electricity, housing, transportation, and food, then evaluated possible correlations with 37 independent demographic variables.

The results of the authors' analysis exhibit different patterns depending on geographic region, but some national trends are evident. One of the most obvious patterns is the relationship between mean HCF and population density. Metropolitan regions tend to consist of a low-HCF urban core surrounded by a ring of much higher consumption in the outlying suburbs. Rural areas farther away from urban centers have low-to-average HCF values. This trend is most extreme in larger cities, which have the lowest HCF of all in areas of particularly high population density. Unfortunately, the particularly extensive suburbanization that occurs in these areas more than counteracts the benefits of relatively low carbon emissions in the city centers. The details of emissions patterns, however, are not as predictable. The composition of total HCF, for example, varies significantly by location. Transportation-related emissions, for one, range from 26% to 42% across different regions, due to differences in details such as average commuting time and vehicle fuel efficiency.

A linear regression analysis found that 92.5% of the variability in HCF across all zip codes can be attributed to six variables, the most important being number of vehicles per household, followed by annual household income, carbon intensity of electricity, and home size (measured in number of rooms). The regression analysis also revealed that, when other variables are controlled for, population density itself is actually not significant. Rather, it influences home size, vehicle ownership, and other factors, which in turn affect HCF. Thus, while population density is effective in representing trends of local home size, transportation, and income, these factors are what directly impact HCF, not density itself.

The results of this research provide a set of valuable guidelines for urban planning. First, they reveal a nationwide relationship between HCF and population density: carbon emissions increase with population density up to a threshold of about 3000 persons per square mile, after which HCF decreases with increasing population density. Similarly, because of this trend and the relatively high HCF levels in large suburbs, high population density should not necessarily be an urban planning objective because it tends to be accompanied by high consumption in outlying areas. Thus, GHG emissions reduction efforts must acknowledge the location-specific impacts of different planning strategies. Given the limited capacity of clean energy initiatives to reduce carbon emissions, consumption-driven strategies are essential, and Jones and Kammen's work adds to the body of crucial planning knowledge.

The Cost-Effectiveness of Southern California Public Transportation Projects

California recently established a carbon cap-and-trade program in the interest of improving air quality and fighting global climate change. The tax revenues generated from this program are used to fund projects that help reduce greenhouse gas (GHG) emissions, but policymakers face the challenge of deciding which projects are best suited for this funding. To inform this decision-making process, Matute and Chester (2015) compared the effectiveness of different current and future public transportation projects to determine which is the most cost-effective, in terms of public dollars spent per ton of carbon dioxide equivalent released. The study compared four projects from Los Angeles County: the Orange Bus Rapid Transport (Orange BRT) line in the San Fernando Valley, the Gold Light Rail Transport (Gold LRT) line that runs from Los Angeles to Pasadena, a bicycle and pedestrian pathway along the Orange BRT line, and the California High Speed Rail (CAHSR) project, a plan being developed to expand high-speed rail throughout the state. All four projects were found to have negative costs per ton of carbon dioxide reduced, indicating that they actually save the public money over time. For a 100-year period, the bicycle pathway was found to be most cost-effective, followed by the Gold LRT, Orange BRT, and CAHSR.

Throughout their paper, the authors stressed the importance of the economic metrics used to assess the effectiveness of each project. They quantified this significance by computing the per-ton cost of reduced emissions according to four different methods. According to each of the first three methods, which considered public costs for initial construction, continued operation, or both, none of the projects could reduce pollution at a cost less than the cap-and-trade rate of $12/metric ton. However, the authors also developed their own cost analysis framework that incorporated into cost assessment the expenses saved by the public due to the new systems—that is, reduced automotive and air travel. When these reductions in private costs were taken into account, all four of the projects analyzed were found to be cost-effective, saving hundreds to thousands of dollars per ton of carbon dioxide equivalent emissions reduced.

Matute and Chester's research paves the way for future assessments of the cost-effectiveness of public transportation projects. At a time when California is considering a wide assortment of public transportation projects to reduce automobile traffic, air pollution, and GHG emissions, comprehensive assessments of the environmental and economic impacts of these programs are needed to help policymakers make better decisions. In their analysis, the authors highlight how much of an impact the economic assessment technique has on study results and the uncertainty in their own findings, while also proposing a novel assessment system that more accurately quantifies the economic impacts of different projects. Their thorough approach will be useful in determining how California can do the most environmental good with the cap-and-trade fees it collects and continue to be a national leader in environmental progress.

Conclusions

The research described in this chapter captures only a fraction of the work being performed to develop strategies for dealing with urban expansion in the near future. Urban development poses a threat to sustainability on both the local and the global scale, but a variety of methods have already been implemented to help offset the negative impacts of cities. Most of these strategies on their own do not do enough to counteract the environmental damage future urbanization is expected to cause; however, recent research does present a number of effective methods to mitigate and adapt to climate change in cities. Reflective and green roofs, proper utilization of brownfields, economic incentives, well-planned public transportation systems, and more can minimize the urban heat island effect and reduce greenhouse gas emissions. Each of these proposed methods may bring unintended consequences that hinder progress, but the proper combination of these emerging techniques should significantly offset the potential harms done by modern urban development.

References Cited

Barau, A.S., Maconachie, R., Ludin, A., Abdulhamid, A., 2015. Urban morphology dynamics and environmental change in Kano, Nigeria. Land Use Policy 42, 307-317.

Baró, F., Chaparro, L., Gómez-Baggethun, E., Langemeyer, J., *et al.* 2014. Contribution of ecosystem services to air quality and climate change mitigation policies: The case of urban forests in Barcelona, Spain. AMBIO 43, 466-479.

Creutzig, F., Baiocchi, G., Bierkandt, R., Pichler, P., Seto, K. C., 2015. Global typology of urban energy use and potentials for an urbanization mitigation wedge. Proceedings of the National Academy of Sciences, published ahead of print.

Georgescu, M., Morefield, P. E., Bierwagen, B. G., Weaver, C. P., 2014. Urban adaptation can roll back warming of emerging megapolitan regions. Proceedings of the National Academy of Sciences 111, 2909-2914.

Jones, C., Kammen, D. M., 2014. Spatial Distribution of U.S. Carbon Footprints Reveals Suburbanization Undermines Greenhouse Gas Benefits of Urban Population Density. Environmental Science & Technology 48, 895-902.

Mathey, J., Robler, S., Banse, J., Lehmann, I., Brauer, A., 2015. Brownfields as an Element of Green Infrastructure for Implementing Ecosystem Services into Urban Areas. Journal of Urban Planning and Development, published ahead of print.

Matute, J.M., Chester, M.V., 2015. Cost-effectiveness of reductions in greenhouse gas emissions from high-speed rail and urban transportation projects in California. Transportation Research Board 94th Annual Meeting, 1-14.

Rotem-Mindali, O., Michael, Y., Helman, D., Lensky, I. M., 2015. The role of local land-use on the urban heat island effect of Tel Aviv as assessed from satellite remote sensing. Applied Geography 56, 145-153.

Anticipating Climate Change Impacts on Human Health

Amelia Hamiter

Preparing for the ways global climate change will alter human health needs is difficult. Both factors and the relationship between them are affected by countless variables such as changes in human populations' ways of living, ecological changes affected by climate, socioeconomic shifts, and more. Despite the high margin of unpredictability, researchers have continued to study how specific climate change scenarios might have specific public health outcomes.

In this chapter I review papers which have modeled health outcomes of climate change in several specific areas of concern which can be categorized in many different ways. Some of these papers are focused on specific diseases, such as malaria, chronic kidney disease, or Chagas disease, and likewise often look at a specific geographic location. A few papers focus on specific environmental factors such as heat waves or water quality and their impact on local public health. Others do not look directly at human health, but study how climate changes affect species that transmit human diseases. Some consider the effects of human behavioral adaption. Still others study the methods of studying climate change, offering feedback on how public health researchers can better project global climate scenarios. Many of these papers fall into more than one of these categories.

Several papers deal with the question of how infectious disease vector populations will be affected by environmental changes, and thus how their patterns of transmission will alter. The vectors studied are small species that transmit diseases to humans. The ways in which environmental changes might affect transmission rate are many; they could affect parasite and pathogen species that interact with the vector, alter the vector's biting rate, change the geographic niche in which the vector population can reside, and subsequently expand or decrease the human populations they affect (Caminade *et al.* 2014, Medone *et al.* 2015). Concerns about how climate change will affect these species drive these studies, but their modeling methods do possess a high degree of uncertainty as they are based on assumptions about how a vector population will react to climate change (Egizi *et al.* 2015).

Socioeconomic factors are important to include in studies of public health and climate change. One concern is that many tropical and subtropical countries are currently considered low- to middle-income, and would struggle economically should global warming decrease the output of their workforce. These also are often nations in which many workers labor in outdoor or unventilated environments where climate warming will increase occupational health hazards. Urbanization also factors into this, with the "urban heat island" effect describing the greater absorption of heat by surfaces found in greater concentrations in cities than in rural areas, making cities hotter (Kjellstrom 2015).

Heat stress might be considered a more direct public health threat of climate change, worsening conditions of those who are already in ill health and inducing adverse health conditions in previously healthy individuals (Kiranmayi *et al.* 2014). Heat waves with high mortality in the early 21st century more drastically brought to attention the threat of global warming. The fallout of extreme heat waves such as the 2003 event in France discussed in one of the papers here has been recorded and analyzed (Lemonsu *et al.* 2014), and that work has been found useful in studies of later heat waves, such as in 2010 in Russia (Shaposhnikov 2015). The urban heat island effect also comes in as a relevant factor increasing urban heat stress.

Other environmental elements such as large bodies of water that affect local human health may in turn be affected by climate changes.

In the Mekong Delta Basin, water quality already contributes to local health conditions in multiple ways, and it is important to consider how climate change will alter that (Phung *et al.* 2015).

While many public health studies project countries' vulnerability to natural impacts based on environmental conditions, countries' ability to adapt is another important element that ought to be taken into consideration when modeling vulnerabilities. A related variable is difference in regional healthcare systems and resources (Suk *et al.* 2014).

The question of how human populations will adapt on a more immediate scale is also very important (Parham *et al.* 2015). As projections of climate change's influence are published and circulated, how will countries, public health organizations, and other parties react? Changes made by human populations as research is being done in this area should be documented, and it should be considered whether these changes will affect the projected impact of climate change.

Modelling Influence of Climate Change on Global Malaria Distribution

Malaria is an infectious disease which has significantly affected global human populations throughout history. Its impact declined greatly throughout the 20th century in many regions due to extensive intervention efforts, but it continues to be found in tropical areas such as parts of Africa. As for other vector-borne diseases, though, global climate change could potentially alter the distribution and "seasonal activity" of malaria. One way to prepare for changes in patterns of malaria outbreaks is to model scenarios of change under different climate outcomes. Fortunately, malaria is one of the few climate-sensitive health events for which more than one research group have done such a study. Since a variety of malaria models of both global and regional scale exist, it is possible to do a "multimodel intercomparison," which results in a more thorough projection of potential climate change effects. Caminade *et al.* (2014) are the first to do this for malaria models. They conclude that the most significant impacts of climate change are likely to happen in the highlands in Africa, and in parts of South America and Southeast Asia, and less likely to happen in more developed regions, most likely because of the socioeconomic factors which enabled these regions to reduce malaria in the 20th century, such as strong public health services. Because the malaria models left out consideration of factors such as human migration and urbanization, technological advances, economic developments, change in land use, changes in vector and disease control, and adaption and evolution of the malaria parasite and vector species, they are dearly uncertain, as the authors note. They also did not account for how other effects of climate change (such as social, economic, and political changes) will have an impact on how human populations of the future will be able to manage disease outbreaks. Considering that human interventions had a significant impact on malaria distribution in the 20th century, it is important to look at such factors in a malaria model. The researchers add that as future biological models become more complex, the differences between them will grow, making it worth questioning whether it would be better to stick to the most simple models instead. They also advise doing further malaria model work at regional or national scales to glean a more accurate picture of populations at risk. The authors conclude that while climate-sensitive alterations in malaria distribution are important, it is "unlikely" that they will be a major player in the net disease outcomes of climate change.

This study was done as a part of the Inter-Sectoral Impact Model Intercomparison Project, a collaboration that develops methods of intercomparison in order to better understanding of impact models. The researchers used five malaria models: LMM_RO, MIASMA, VECTRI, UMEA, and MARA, all of which were developed for different regions and with different objectives, and had different parameters and designs. This intercomparison used climate outputs from five global climate models (CGMs), running each under four different emissions scenarios (these were Representative Concentration Pathways (RCPs) developed for the fifth Intergovernmental Panel on Climate Change (IPCC) assessment report). Using these tools and a single population projection, researchers evaluated climate suitability—how the climate may become more suitable for malaria transmission—and populations at risk. Their findings indicated net increase in both, but with large uncertainty due to the fact that the malaria models did not include possible effects of many factors affecting malaria distribution besides climate change.

Modeling Variables of Climate Change Impact on Distribution of Vectors of Chagas Disease

Concerned with epidemiological implications, Medone *et al.* (2015) assess bioclimatic variables that are the most influential for distribution of *Rhodnius prolixus* and *Triatoma infestans*. These are both vectors of *Trypanosoma cruzi*, a parasite that spreads Chagas disease in Latin America. To project the impact that climate change will have on the distribution of these species in the coming years, this study models suitability—the likelihood of finding a niche of environmental conditions suitable for maintaining a population without need for immigration—for the species' presence in rural populations in Venezuela and Argentina through the year 2050. Temperature-related variables were found to be the most dominant contributing bioclimatic variables for both species, with some precipitation variables also coming into play. The suitability findings predict that *R. prolixus* in Venezuela will see future expansion to new areas, but possible decrease in some areas where it currently is distributed. Meanwhile, the geographical range of *T. infestans* is projected to decrease. The results predict a decreasing trend in the number of new human cases of *Tr. cruzi* infection per year between now and 2050, but the authors point out that the size of rural populations in the studied areas has been declining in past years, and if it continues to do so, a decrease in infection incidence will no longer directly imply a decline in transmission risk since the human population has decreased. The authors concede that their modeling methods were not able to account for socio-environmental or socio-economic factors, which would affect conditions for future disease transmission. They also concede that they had limited datasets, and that predictions for future conditions are never certain. With these limitations, they advise interpreting their results not as indicators for short-term vector control planning but as indicators of possible longer-term trends. They recommend further research to specify future *Tr. cruzi* transmission risk.

This study uses ecological niche modeling to predict suitability for *R. prolixus*, a tropical species, and *T. infestans*, a temperate species. They use entomo-epidemiological information from a wide range of sources and databases, and use bioclimatic data layers from World-Clim. To consider future climate scenarios and analyze different bio-climatic variables they ran a principal component analysis (PCA), and used MaxEnt (Maximum Entropy Modeling) to model suitability for the species at current times and in 2050. To convert climate suitability results to transmission risk, the researchers used force of infection (FOI), which measures the rate at which susceptible individuals become infected. They used both a one-step and a two-step approach for FOI, and found that the one-step method, which tracks transmission risk in terms of human age class, worked better in Argentina while the two-step model, which considers transmission risk based on household infestation and vector density per house, was more beneficial in Venezuela. The most important bioclimatic factor for climatic suitability was minimum temperature of the coldest month for *R. prolixus*, and mean temperature of the coldest quarter for *T. infestans*. Other temperature- and precipitation- related factors were found to be important for both species as well.

Studying Vector Habitat Range Expansion as an Analogy for Species Evolution in Response to Climate Change

Many studies model future climate change and its probable impact on disease vectors over long periods of time; these models often do not take into account possible evolution of the vector, assuming that climate changes will occur faster than the species can evolve. However, Egizi *et al.* (2015) consider it essential to determine whether evolution is a viable response of vectors to climate change, as that could significantly influence the effects of climate change. Predicting future patterns of evolutionary response of vectors to climate change is difficult. However, a proxy can be found in evaluating the evolutionary patterns of invasive vectors that have expanded their habitat range to thrive in different environments. Rather than encountering a temporal "climate change," such a population encounters a geographic "climate change." Whether it evolves accordingly and the timeframe in which it does so provide insight into how a similar species might evolve in response to temporal climate change. Studying invasive populations' evolution is complicated by movement such as reinvasions. However, studying predictive factors (genetic diversity, viable mutation rate, and selection pressure in different vector populations as well as movement between populations) still allows researchers to evaluate the species' evolution in response to new climes. Egizi *et al.* carry out such an analysis by evaluating changes between populations of *Aedes japonicus japonicus*, a mosquito that has invaded new habitats across the globe within a decade. Their results showed that significant genetic distinctions did arise between populations living in different climates. While these results do not say if these developments increase fitness, and thus are *adaptive*, they do show that it is not implausible for a novel environment to drive population evolution. This implies that a vector could indeed evolve at a rate comparable to changes in its climatic environment, and a climate change impact model that assumes otherwise neglects an important factor. This issue is especially significant in studies related to disease vectors, as evolved changes in a vector genome could affect phenotypes important to its transmission of the disease to humans.

Ae. j. japonicus is an ideal species for this study. It shares traits with close relatives that are known to be critical disease vectors, and detailed data are available recording genetic variations in its invasive populations. *Ae. j. japonicus* is native to northern Japan and the Korean peninsula, but populations spread to Europe and North America in the early 2000s despite variances between those habitats and its native climate. This study examines the genetic variation of *Ae. j. japonicus* populations that have been established in different climes in the state of Virginia and on the island of Hawai'i. For both regions, the genetic distinctions that developed between these populations in the short 7–10 year timespan correlate with the temperature variance of the populations' respective habitats. The nature of genetic change differs between the two study locations—allelic diversity (genetic variety) characterizes difference between populations on the island of Hawai'i, while genetic distance (genetic divergence) characterizes difference between populations in Virginia. The climatic and topographical traits of these regions reasonably account for this divergence—Hawai'i populations are more isolated and thus have less gene flow than populations in Virginia, which are more connected to thriving populations in nearby states. Hawai'i also has a steeper temperature gradient, yet less annual temperature fluctuation than Virginia, both factors that may influence selection pressure and number of generations per year for *Ae. j. japonicus*, thus the diverging forms of genetic variation.

The authors use these findings of temperature correlating with evolution in invasive species analogously to imply that a vector population could plausibly evolve in response to climate change. While many other factors complicate the both species adaption and environmental changes, it is important for future researchers to understand that both phenomena can occur together. Continuing the analogy, the authors note that while reinvasion is a complicating factor for invasive population evolution, climate change's impact on other species (particularly humans, pathogens, and parasites) adds to climate change's implications for vector populations.

Occupational Health Hazards and Consequent Economic Losses Due to Workplace Heat Exposure

Kjellstrom *et al.* (2015) study how warming temperatures due to climate change may create an occupational health hazard in tropical and subtropical countries that have a significant workforce employed in jobs in hot environments, such as physical jobs which must be done outdoors or in indoor spaces that lack efficient cooling systems such as many factories. This problem is exacerbated by the high humidity of these countries, which reduces the effectiveness of sweating in cooling the body. To avoid excessive heat stress, workers must not work during the hottest hours of the day, which increase in the hottest days of the year. This loss of work hours can affect the respective gross domestic products (GDPs) of the countries affected by this, which are often already low- to middle-income. Preventative actions can be taken such as development of coolant systems where possible as well as occupational health advisories, adjusted work hours, and other changes such as increased access to drinking water and education about symptoms of heat strain and heat stroke in the workplace. However, these strategies are limited, and also hold little hope for cutting economic losses. Global action against climate change is the most effective action to take against this situation.

Little assessment has heretofore been done on the effects of heat exposure raised by climate change on people who work in hot climates and on consequent economic consequences. Once surrounding air temperature rises above 37°C, or 98°F, heat from the air transfers to the body, and the body must be cooled by the evaporation of sweat. In high humidity, however, sweating becomes less effective at reducing body heat. Heat stress thus becomes an especially prevalent problem for workers in the humid and prevalently hot weather seasons of tropical and subtropical countries. Workers in these occupations often cannot work during the hottest times of the day and in the hottest days of the year; they must rest more and work more slowly to cope with heat. Physical acclimatization to heat varies individually and is limited to evaporation of sweat from the skin, which has little effectiveness in high humidity.

To further assess current and future climate change impact on local heat exposures for working people, the global Hothaps program has been developed. To evaluate heat data, this program used WBGT (Wet Bulb Globe Temperature), a common heat exposure index that in one value combines temperature, humidity, wind speed, and heat radiation. Only indoor values were evaluated (with the knowledge that the outdoor values tend to be a few degrees higher). Data from weather stations around the world were used to calculate time trends, which communicate trends in monthly or annual averages and trends in number of days above a specific threshold in a year. The

"Hothaps-Soft" software developed by this program uses WBGT as heat stress index for workplace conditions and has been made freely available at the site www.ClimateCHIP.org along with the weather station data. This paper shows these calculated time trends graphed for the month of May in Kuala Lumpur, Malaysia.

Additionally, the heat situation of different parts of the world was mapped using WBGT of different months and years calculated from data from the Climate Research Unit at University of East Anglia, UK, and future model data were also included. The paper also shares a heat map for regions of India that indicates reduction in hours of workable daylight in different weather seasons. The authors note, however, that the climate data used most likely mainly represents less urbanized areas, yet urban areas have higher ambient temperature than rural areas due to the "urban heat island" (UHI) effect due to absorption of heat into concrete, asphalt, and other surfaces common in such areas. Hothaps-Soft can compare climate data and estimated WBGT levels at central city weather stations with data from airports to estimate heating trends in urban areas.

The workplaces in which heat strain is a significant issue are difficult to cool, even if they are located indoors. Installing air conditioning in urban workplaces, for example, is highly contested as a sustainable solution since large-scale air conditioning use actually heats local outdoor air and increases electricity demand, and can create a huge energy demand for large cities.

These calculated losses in hours of work were used to estimate economic losses. The authors' model of work hour loss due to heat evaluated the proportion of national workforces in jobs of varying levels of physical demand and heat exposure along with population estimates for different geographic regions. Future GDP estimates based on international projections were then multiplied with the calculated cumulated annual work hour losses, presented as percentages of daylight work hours. These losses were calculated for different regions for "baseline" (1960–1989) and for future periods (2030 and 2050), with consideration of potential socioeconomic changes and changes of workforce distribution into different industries. The paper includes several tables that show a selection of heat and economics data for varying countries. Some of the countries shown are specified as belonging to the Climate Vulnerable Forum (CVF), an organization of 20 low- and middle-income countries that consider themselves threaten by a climate change risk that has been caused largely by other entities. Compared in the tables to countries not in this category, it is seen that the CVF countries and other low- and middle-income countries experience more days of serious heat per year, greater heat impact change, and greater economic loss estimates, than the non-CVF and high-income countries. Estimated annual losses reach the multibillion dollar levels for several of these low- and middle-income countries, showing that climate change can significantly hamper these countries if not mitigated.

Increase in Hot Environments by Climate Change Can Increase Chronic Kidney Disease Rates

Most studies on the relation between heat stress and renal dysfunction have focused on developed Western countries, but different climates and socioeconomic factors could have implications for that relation in other regions. Kiranmayi *et al.* (2014) investigate the effects of heat and developments of chronic kidney disease in low-income areas of India, a country that has different environmental conditions from the nations in which this relation has previously been studied. They observe that extreme exposure to hot weather heightens the risk of kidney injury in healthy individuals and speeds progression towards chronic kidney disease (CKD) in individuals with preexisting kidney conditions or other conditions that make them vulnerable to renal dysfunction.

Nephrons are the units of the kidney which perform its main function of filtering the body's blood in order to absorb needed substances and excrete the rest as urine. Chronic kidney disease can result from either structural kidney damage or from a decrease in glomerular filtration rate (GFR)—that is, the nephrons' filtering capacity. GFR is a key quantification that assists in early detection of renal impairment, and CKD is classified into five stages based on patients'

GFR. A marker of GFR is serum creatinine, which has previously been found to rise in the majority of patients affected by heat stress. In this study, blood urea and serum creatinine levels were measured in a range of individuals. The study used a control group of 123 healthy individuals who were free of kidney disease features and had no diseases or drug intake histories likely to change their blood urea and serum creatinine level parameters. One hundred and ninety-eight cases of patients with renal insufficiency and CKD evidence comprised the test group; patients with CKD signs were classified by their types of treatment – there was a non dialysis subgroup and a hemodialysis subgroup. This study found that blood urea and serum creatinine levels increased in patients with CKD compared to those in the control group, indicating decline of GFR in those patients.

Increases in kidney dysfunction occurred at the hotter times of the year—the cases studied had registered within the months of March and May, when their regions saw rises in temperature and humidity. Most of the registered patients in the study were agriculture, construction, or industrial labor workers of low income standing. These workers' physical labor done in hot environmental conditions during the summer could easily lead to hyperthermia, the elevation of body temperature that results when a body absorbs more heat than it dissipates. Hyperthermia can induce volume depletion (the loss of extracellular fluids in the kidney), which leads to repeated subclinical kidney injury in previously healthy individuals which in turn can progress to chronic kidney disease. The authors note that physical labor done by these patients also often caused muscle damage, which could cause subclinical kidney damage that would progress to CKD. Heat and dehydration could also injure the kidney for these workers by changing blood perfusion, thus making it difficult for the kidney to get enough blood. Sweating is the body's main way of dissipating heat, but if it occurs continuously without rehydration the consequent water loss can also lead to volume depletion for the kidneys. In humid climates such as parts of India, evaporation of sweat is decreased and thus it is not as effective in cooling the body; this makes it easier for the body to absorb heat and undergo hyperthermia.

The study also observed decline in GFR in individuals of advanced age. Aspects of aging in the body overall as well as in the kidney make elderly individuals more vulnerable both to the effects of heat stress and to renal failure.

The researchers mention other demographic vulnerabilities to kidney disease observed by other studies. Obese individuals and those suffering from diabetes mellitus also have biological factors that make them more susceptible to heat exposure's effect on the kidneys; adipose tissue decreases water content and is a less effective surface area, while diabetes mellitus is a nephrotoxic condition that causes progressive damage to the kidneys. Therapeutic drugs taken for various conditions that inhibit thermoregulation also make their recipients more at risk to the consequences of heat strain for renal function. Thus a range of factors makes specific populations especially vulnerable to the effects of heat strain.

The authors acknowledge that their study had sample size and regional limitations, with most of its patients being from the northeast part of Andhra Pradesh in India. The authors conclude that the best action to take is to implement health education programs raising awareness of heat-related illnesses and their relation to climate change.

Projecting the Frequency of Heat Waves in the 21st Century over the Paris Basin

The high mortality of the 2003 heat wave in France, and its particularly severe impact in the Paris basin, has drawn attention to the importance of considering heat wave occurrences of the future. Evaluating heat waves in the Paris region from 1951-2009 and using several climate change and emissions scenarios to model future heat wave possibilities, Lemonsu *et al.* (2014) predict that the frequency of heat wave occurrences in the target area will increase systematically with time and global warming, and that the durations of these heat waves will grow.

Utilizing the approaches of various climatic studies and the PNC (Plan National Canicule, a national operation warning system set up in the aftermath of the 2003 heat wave), the researchers created a definition of a heat wave that would work with the constraints of climatic projections and took into account public health impacts. The researchers use a reference historical time series to evaluate heat wave occurrences of the past and compare those data with projections of future heat wave frequency. They used daily minimum and maximum air temperature records from various meteorological stations throughout the region, with the longest station records dating from 1945 and the shortest dating from 1974, as well as data from the Météo-France archives, which document main heat wave events in France since 1950. With this data, researchers could extract occurrences observed in the past that they defined as heat waves.

The researchers extracted future heat wave projections using climate and emissions scenarios. Two databases, ARPEGE-Climate-v4 and ENSEMBLES, provided daily minimum and maximum temperature time series from 1950 to 2099, and emission scenarios were applied from 2001. With these time series simulations, projections for different scenarios of future heat waves could be made and compared with past heat waves, allowing researchers to predict the frequency and duration of heat waves over the next century under the various climate and emission scenarios. Similar methods were applied to six other locations in France to ensure the quality of the study methods.

The results of this study stress the need to employ mitigation and adaption strategies to prepare cities for events of extreme heat. The urban heat island effect makes taking such actions even more imperative.

Evaluating the Need for Research on the Effects of Climate Change on Water Quality and, in turn, Health Conditions in the Mekong Delta Basin

Studies of climate conditions' relationship with water quality and public health have been done in developed areas, but not for bodies of water such as the Mekong Delta Basin, which runs through several countries in Southeast Asia. Phung *et al.* (2015) perform an analytical review of research that has been done on these relations in the Mekong Delta Basin in order to compile findings thus far and to see what areas of concern should be given more study. They found that research had been done in the area linking health concerns to both extreme weather events and regular seasons, but not in a framework that thoroughly considered climate change. Likewise, findings on water quality rarely included analysis of climate change factors.

Researchers searched several major databases such as PubMed and GoogleScholar as well as references from relevant publications, having selected specific keywords and criteria, and compiled data from papers focused on the Mekong Delta Basin. Using findings from studies on climate change, disease, and water basins in other regions, as well as Mekong-specific findings, they evaluated water-related health concerns that might be most pertinent in climate change events. From there, they considered what gaps in research existed for the Mekong region. The authors acknowledge that their review has several limitations. Research on climate change and health in the Mekong area has not been extensive, and the papers they find have a wide variety of methods and quality. Generally, climate and health can be very indirectly linked and are affected by a wide array of other factors, making it difficult to pin down the exact influence of climate change.

Some studies in the Mekong Delta region evidenced connection between water quality and climatic factors on a preliminary level, but would need to go further to examine how the effect of climate factors varying over time. The authors emphasize that chemical contamination is a subject that could definitely be given more attention. Studies in other areas have found that this issue is affected by changes in climate factors such as temperature and moisture alterations, and its effect on water quality has many implications for local human health. Climate change's impact on groundwater in the Mekong Delta region, which could provide information important for developing adaption strategies, has also been neglected.

Studies of disease near the delta basin have linked outbreaks of diarrhea, typhoid fever, mosquito-borne diseases, and other water-related diseases to climate events such as flooding and the wet season.

However, little work has examined the factor of changes in climate over time. Health conditions that result from water contamination and that would be affected by climate change also have not been studied in the Mekong Delta region.

Tracking Vulnerabilities to the Risks of Infectious Disease Transmission due to Climate Change in Europe

Suk *et al.* (2014) examine vulnerability to infectious disease transmission changes as a measurement of both the impact of climate change on transmission in a region and the region's ability to respond (described here as adaptive capacity). This concept of vulnerability differs from that used in most public health practices, which generally do not take adaptive capacity as a component of vulnerability. Indeed, the authors note that the health sector has produced little research that examines infectious disease transmission due to climate change or the effects of different socioeconomic development pathways in studies of vulnerability. Thus, they take on the task of creating a quantitative indicator to measure regional vulnerabilities that combines all of these factors. Their projections assess which regions are projected to undergo climate changes more significant than their adaptive capacities, and thus are particularly vulnerable to infectious disease transmission changes. They evaluate that some of these high vulnerabilities are driven by low adaptive capacities, while others have high adaptive capacities yet face enough projected climate change that they are still highly vulnerable. The researchers recommend that the next steps forward are to carry out more disease-specific and more detailed health indicators of vulnerability studies.

Researchers used the European Observation Network, Territorial Development and Cohesion (EPSON) approach to modelling vulnerability, which measures vulnerability as the reciprocal relationship between climate change impact and adaptive capacity. Their index parameters projected vulnerability for the years 2035 and 2055 for subnational regions of the European Union Member States. Their index is general rather than disease-specific, with the intent that it could in the future be combined with data regarding specific infectious diseases to asses more specific indicators and risks.

Temperature and precipitation were the two variables for measuring climate change impact. For adaptive capacity, the index used the EPSON dataset, which takes into account factors such as public health and healthcare infrastructure, economic resources, governance, etc. The resulting index has plenty of limitations, as noted by the authors. Two main limitations they mentioned are that future-oriented models cannot be well validated, and that regional variability and varying healthcare system structures are difficult to account for.

The index results project that the highest adaptive capacities (and lowest baseline vulnerabilities) are in Scandinavia, southeast England, and central Europe, while the lowest adaptive capacities (and highest baseline vulnerabilities) are in southern and eastern Europe. The regions facing the largest changes in temperature and/or precipitation are expected to face the largest climate-related infections disease problems. In some regions, the adaptive capacity is able to mitigate climate change impacts to some extent, while in others, lower adaptive capacities lead to higher vulnerability index rankings even if the impact index is not as high. Some regions in the Iberian peninsula, the UK and Ireland, and southern France, rank in higher quintile ranges for both impact and vulnerability indices, indicating that regardless of the level of their respective adaptive capacities, they are not high enough to mitigate the expected high impact of climate change.

Conclusions

This survey of recent work done in these topics I hope conveys an idea of the diverse ways in which climate change has the potential to affect global health needs. Even if the specifics are impossible to predict, human health concerns will change as regional climates do, and both local and global public health workers must be prepared to respond. While countless variables make it almost impossible for researchers to dictate the best mode of preparation for the future, gaining an idea of the scope of possible consequences gives epidemiologists, policymakers, and other involved parties a better awareness of the diversity and unpredictability of scenarios to which they will need to react.

References Cited

Caminade, C., Kovats, S., Rocklov, J., Tompkins, A.M., *et al.* , 2014. Impact of climate change on global malaria distribution. PNAS 111, 3286-3291.

Egizi, A., Fefferman, N.H., Fonseca, D.M., 2015. Evidence that implicit assumptions of 'no evolution' of disease vectors in changing environments can be violated on a rapid time-scale.Philosophical Transactions B 370, 10.1098/rstb.2014.0136.

Kiranmayi, P., Raju, D.S.S.K., Vijaya Rachel, K., 2014. Climate Change and Chronic Kidney Disease. Asian J. of Pharmaceutical and Clinical Research 7, 53-57.

Kjellstrom, T., Meng, M., 2015. Impact of Climate Conditions on Occupational Health and Related Economic Losses: A New Feature of Global and Urban Health in the Context of Climate Change. Asia-Pac. J. of Public Health 10.1177/1010539514568711.

Lemonsu, A., Beaulant, A.L., Somot, S., Masson, V., 2014. Evolution of heat wave occurrence over the Paris basin (France) in the 21st Century. Climate Research 61, 75-91.

Medone, P., Ceccarelli, S., Parham, P.E, Figuera, A., Rabinovich, J.E., 2015. The impact of climate change on the geographical distribution of two vectors of Chagas disease: implications for the force of infection. Philosophical Transactions B 370, 10.1098/rstb.2013.0560.

Parham, P.E., Waldock, J., Christophides, G.K., Hemming, D., *et al.* 2015. Climate, environmental and socio-economic change: weighing up the balance in vector-borne disease transmission. Philosophical Transactions B 370: 10.1098/rstb.2013.0551.

Phung, D., Huang, C., Rutherford, S., Chu, C., Wang, X., Nguyen, M., 2015. Climate Change, Water Quality, and Water-Related

Diseases in the Mekong Delta Basin: A Systematic Review. Asia-Pacific Journal of Public Health 10.1177/1010539514565448.

Shaposhnikov, D., 2015. Long-term impact of Moscow heat wave and wildfires on mortality. Epidemiology 26, e21-e22.

Suk, J.E., Ebi, K.L., Vose, D., Wint, W., *et al.* 2014. Indicators for Tracking European Vulnerabilities to the Risks of Infections Disease Transmission due to Climate Change. Int. J. Environ. Res. Public Health 11, 2218-2235.

Climate Change and Human Health

Allison Hu

Today, worldwide, there is an apparent increase in many infectious diseases, including newly circulating ones. This reflects the combined impacts of rapid demographic, environmental, social, technological, and other changes in overall living conditions and customs. This chapter focuses largely on the impacts of climate change on infectious disease occurrence and other human health factors such as diet, winter mortality, and solar ultraviolet radiation.

Many prevalent human infectious, including infectious gastroenteritis, tuberculosis, *Mycoplasma pneumonia*, and waterborne and water-related infectious diseases are climate sensitive. Climate restricts where an infection can occur because it limits the distribution of other species vectors that are required for disease transmission. Because many of these diseases remain a major health risk in many parts of the world, gaining a better understanding of their sensitivity to climate could aid in the development of a reliable climate-based prediction system for epidemics, which could lead to improvements in the current disease control program in countries where these diseases are an endemic.

Furthermore, climate change not only impacts the transmission of infectious diseases but also other aspects of human health such as winter mortality, drought and fires creating health-hazardous agents, diet, and UV radiation. Developing a better understanding of the dynamic relationship between climate variation and these other factors can create vital preventative measures from enforcing fire control legislation to mitigating fire impacts on ecosystems and on human health, to adopting alternative diets for both global environmental and public health benefits.

Effect of Non-Stationary Climate on Infectious Gastroenteritis Transmission in Japan

Infectious gastroenteritis, otherwise known as the stomach flu, is a medical condition from inflammation of the gastrointestinal tract that involves both the stomach and the small intestine, causing a combination of diarrhea, vomiting, and abdominal pain and cramping. This common disease contributes significantly to the 1 billion episodes of diarrhea and 3 million deaths in children under 5 years of age per year, and is the fifth-leading cause of death worldwide (Onozuka *et al.* 2014). The transmission of infectious gastroenteritis is rather complex, involving both host and environmental factors.

Environmental factors such as temperature, humidity, and rainfall are widely considered as important factors in the spread and seasonality of infectious gastroenteritis. Additionally, several studies reported that the El Niño-Southern Oscillation (ENSO) and Indian Ocean Dipole (IOD) play important roles in the transmission of infectious diseases – including dengue, malaria, and cholera (Onozuka *et al.*)

The ENSO is a naturally occurring phenomenon that involves fluctuating ocean temperatures in the equatorial Pacific. This phenomenon is the most dominant force causing variations in regional climate patterns, which affects weather conditions such as temperature, rainfall, wind speed and direction, and storm tracking throughout the world. The pattern generally varies from region to region and fluctuates between two states: warmer than normal central and eastern equatorial Pacific SSTs (El Niño) and cooler than normal central and eastern equatorial Pacific SSTs (La Niña).

The IOD is a coupled ocean and atmosphere phenomenon in the equatorial Indian Ocean that affects the climate of Australia and other countries that surround the Indian Ocean basin. A positive IOD period is characterized by cooler than normal water in the tropical eastern Indian Ocean and warmer than normal water in the tropical western Indian Ocean. A positive IOD SST pattern has been shown to be associated with a decrease in rainfall over parts of central and southern Australia. Conversely, a negative IOD period is characterized by warmer than normal water in the tropical eastern Indian Ocean and cooler than normal water in the tropical western Indian Ocean. A negative IOD SST pattern has been shown to be associated with an increase in rainfall over parts of southern Australia. Moreover, the World Health Organization (WHO) quantified the impact of global warming on diarrhea, and reported that warming by 1°C was associated with a 5% increase in diarrhea. Although regional differences and contrasting effects of temperature on different kinds of diarrhea are evident, few studies have examined the non-stationary relationships between global climatic variability and infectious gastroenteritis.

Onozuka therefore explored the time-varying relationship between climate variation and monthly incidence of infectious gastroenteritis between 2000 and 2012 in Fukuoka, Japan using cross-wavelet coherency analysis to assess the pattern of associations between indices for the Indian Ocean Dipole (IOD) and the El Niño Southern Oscillation (ENSO). Wavelet analysis is useful in the investigation of non-stationary associations using time series data as it can measure associations between two time-series at any frequency band and time-window period. This analysis has been used to determine whether the presence of a particular periodic cycle at a given time in disease incidence corresponds to the presence of the same periodical cycle at the same time in an exposure covariate. Wavelet analyses have also previously been used to analyze the transmission of infectious diseases. Therefore, a better understanding of the sensitivity of these analyses to climate variability can potentially contribute to developing a reliable climate-based prediction system for gastroenteritis epidemics. Onozuka was able to report the first ever quantification of the time-varying impact of climatic factors on the number of infectious gastroenteritis cases using cross-wavelet analysis.

Analysis demonstrated that infectious gastroenteritis cases were non-stationary and significantly associated with the IOD and ENSO for a period of approximately one to two years. Results therefore not only demonstrated the positive correlation between climate change and the spread of infectious gastroenteritis but also the importance of global climate factors such as IOD and ENSO on the transmission of infectious gastroenteritis. Lastly, Onozuka found that the incidence of infectious gastroenteritis is strongly associated with local weather factors such as temperature, relative humidity, and rainfall, with coherent cycles throughout the year. Furthermore, relative humidity and rainfall also affect water and sanitation infrastructure and the number of pathogens, and might impact the replication rate of certain bacterial and viral pathogens that contribute to the "host factor" of the disease.

Results of this study have practical implications for public health officials. Elucidation of the relationship between climate variability and infectious gastroenteritis transmission is important for disease control and prevention. These findings may be beneficial in helping public health officials predict epidemics and prepare for the effects of climatic change on infectious gastroenteritis through the implementation of preventative public health measures. Early warning systems for epidemics of infectious gastroenteritis should consider non-stationary, and possibly non-linear patterns of association between climatic factors and infectious gastroenteritis cases.

Extreme Temperatures Increase the Incidence of Tuberculosis in Japan

Tuberculosis (TB) is a major global public health problem—it affects millions of people annually and ranks as the second leading cause of death from an infectious disease worldwide (World Health Organization (WHO) 2013) (Onozuka *et al.* 2014). The WHO estimated that there were 8.6 million new TB cases and 1.3 million TB deaths in 2012. The worldwide TB incidence rates peaked in 2004 and have decreased at a rate of less than 1% per year since then. Thus, the overall worldwide burden continues to rise as a result of the rapid growth of the world population. TB is a leading cause of death in people in the most economically productive age groups. Furthermore, with growing concerns about global climate change, many studies have focused on associations between weather variability and the fluctuations of infectious diseases and have suggested that weather factors play an important role in their incidence, indicating the possibility of multiple functional pathways.

Although seasonal variation in the incidence of tuberculosis (TB) has been widely assumed, few studies have investigated the association between extreme temperatures and the incidence of TB. Therefore, Onozuka *et al.* examined the possible relationship of extreme temperatures with the incidence of TB cases using surveillance data collected in Fukuoka, Japan, from 2008 to 2012. Time-series analyses was used to assess the possible relationship of extreme temperatures with TB incident cases, adjusting for seasonal and interannual variation. Analysis revealed that the occurrence of extreme heat temperature events resulted in a significant increase in the number of TB cases and the occurrence of extreme cold temperature events resulted in a twenty-three percent increase.

Onozuka *et al.* speculate that extreme temperatures are associated with behavioral patterns that lead to increased contact among people, thereby facilitating the spread of TB infection. Extreme temperatures are also associated with patients presenting for evaluation, rather than being associated with TB incidence. Therefore, it is possible that the explanation for the association observed is that patients are less comfortable when temperatures are high and low, causing them them to seek medical attention. Additionally, vitamin D is an important determinant of adaptive and innate immunity.

There are a few limitations to this study. First, it is possible that TB surveillance data might not capture all of the cases in a community. This under-reporting of infections can occur anywhere in the reporting chain, from the initial decision of a patient to not seek health care to the failure to record cases in the disease registry, due to the mildness or lack of symptoms. Second, TB can often remain undiagnosed for some time if a patient does not seek care or if a clinician does not immediately recognize TB. In these cases, the actual onset of disease would have been prior to the week the case was reported. Lastly, a short period of time was analyzed (5 years) so a longer study period can better determine seasonal patterns of TB.

This study provides quantitative evidence that the number of TB cases increased significantly with extreme heat and cold temperatures. The results may help public health officials predict extreme temperature-related TB incidence and prepare for the implementation of preventive public health interventions (Onozuka *et al.*).

Climate Variability and Nonstationary Dynamics of Mycoplasma Pneumoniae Pneumonia in Japan

Mycoplasma pneumoniae (M. pneumoniae) is a key cause of upper and lower respiratory tract infection and disease in humans, especially in children. This bacterium is estimated to be responsible for 15%–20% of all cases of community-acquired pneumonia (CAP) and up to 40% of cases of CAP among children. More than one-third of these childhood cases of CAP require hospitalization. Furthermore, recent studies have demonstrated that epidemics of *M. pneumoniae* increased significantly with increased average temperature and relative humidity. It is unclear, however, whether elements of the local climate are relevant to *M. pneumoniae* pneumonia transmission have stationary characteristics of climate factors on their dynamics over different time scales. Therefore, Onozuka *et al.* (2014) examined the interannual cycles of *M. pneumoniae* and the potential clinical impacts of these events of CAP.

Onozuka *et al.* performed a cross-wavelet coherency analysis, which is useful in analyzing epidemiological time series, to identify non-stationary associations between disease dynamics and exposure covariates. Wavelet analysis has been particularly used in determining whether the presence of a particular periodic cycle at a given time in the incidence of a disease corresponds to the presence of the same periodic cycle at the same time in an exposure covariate. Onozuke *et al.* assess the patterns of association between monthly *M. pneumoniae* cases in Fukuoka, Japan from 2000 to 2012 and indicators for the Indian Ocean Dipole (IOD) and El Niño Southern Oscillation (ENSO). This is the first report to quantify the time-varying impact of climate factors on the number of *M. pneumoniae* pneumonia cases using cross-wavelet analysis.

Results demonstrated that monthly *M. pneumoniae* cases were strongly associated with the dynamics of both the IOD and ENSO for the one to two year periodic mode from 2005 to 2007 and 2010 to 2011. This association was non-stationary (not consistent over time) and appeared to have a major influence on the synchrony of *M. pneumoniae* epidemics. Consistent with this, an epidemiological study has suggested that El Niño events are associated with viral pneumonia hospitalizations and a more recent study suggested that warming of the Indian Ocean relative to the Pacific might play an important role in modulating Pacific climate changes in the 20th and 21st centuries. These findings suggest that there should be consideration of non-stationary, potentially non-linear, patterns of association between *M. pneumoniae* cases and climatic factors in early warning systems.

Onozuka *et al.* also found that the dynamics of *M. pneumoniae* pneumonia cases are strongly associated with local climate variables, temperature, relative humidity, and rainfall, with coherent cycles of around one to two years in 2004 to 2007. These results are consistent with the previous study mentioned earlier that higher temperature and relative humidity could potentially influence the increase in *M. pneumoniae* pneumonia cases.

With regard to the practical implications of these findings, understanding the effects of climate factors on the epidemiology of infectious diseases is valuable for the planning of health services. Because *M. pneumoniae* pneumonia remains a major health risk in many parts of the world, to gain a better understanding of its sensitivity to climate could aid in the development of a reliable climate-based prediction system for *M. pneumoniae* epidemics, which could lead to improvements in the current disease control program in countries where *M. pneumoniae* is an endemic.

Projected Delays in Reducing Waterborne and Water-Related Infectious Diseases in China Under Climate Change

In 2010, infectious disease due to unsafe water, sanitation, and hygiene (WSH) were estimated to be responsible for 337,000 deaths globally and the loss of over 21 million disability-adjusted life years (DALYs) (Hodges *et al.* 2014). These WSH-attributable diseases include soil-transmitted helminth infections, schistosomiasis, diarrhoeal diseases, and vector-borne diseases such as malaria, dengue fever, and Japanese encephalitis. In China, the WSH-attributable disease burden is concentrated in low-income areas and in young children. Cases not leading to morbidity and mortality, these diseases can causes malnutrition, stunting, impaired school performance, immunodeficiency, and impaired cognitive functioning which can hinder economic growth and development at a population level (Hodges *et al.*). Furthermore, there are studies associating certain diseases with key environmental variables that are responsive to changes in climate, such as temperature, precipitation, and relative humidity. Temperatures changes can influence the replication rate and survival of pathogens and vectors in the environment and impact transmission. Heavy precipitation can overwhelm existing water and sanitation systems, therefore mobilizing pathogens, while drought conditions can increase pathogen exposure by limiting the water available for hygiene and forcing populations to resort to the use of contaminated water supplies.

Although China has taken tremendous strides in reducing these WSH-attributable diseases through decade-long improvements in water supply and sanitation, with the Millennium Development Goal (MDG) for drinking water met in 2009, the MDG for water sanitation has been more difficult to attain and is particularly more vulnerable to risks associated with the impacts of climate change on water supply and quality. China is projected to experience large changes in its climate this century—an increase of 1.8–5.8 ⓧC in temperature compared to the average global temperature increases of 0.3–4.8°C.

Because certain infectious diseases are sensitive to changes in both climate and WSH conditions, Hodges *et al.* projected impacts of climate change on WSH-attributable diseases in China in 2020 and 2030 by coupling estimates of the temperature sensitivity of diarrhoeal diseases and three vector-borne diseases, temperature projections from global climate models, WSH-infrastructure development scenarios, and projected demographic changes. This study is the first assessment of the impact of climate change on WSH-attributable diseases across China.

Although China's demographic and epidemiologic transitions, as well as changes to water and sanitation infrastructure, are expected to rapidly decrease the burden of infectious disease, analysis by Hodges *et al.* demonstrate that climate change will blunt this progress, even while controlling for demographic changes, infrastructure development and urbanization.

In the presence of climate change, China loses on the order of 1–2 years of progress in reducing the burden of diarrhoeal and other WSH-attributable diseases; an additional 15–21 months of continued infrastructure investment, urbanization and demographic shifts will be required to achieve the decreased disease burden projected in the absence of climate change. This projected development delay can be interpreted as the additional years of continued investment in infrastructure and personnel required in the presence of climate change to achieve the disease burden anticipated without climate change. Likewise, results further demonstrated that assuming high emissions (representative concentration pathway (RCP) 8.5) and a maintenance development path, this development delay increases to over 5 years by 2030. Under the most aggressive water and sanitation development path, the development delay by 2030 is 13 months under RCP 2.6 and 18 months under RCP 8.5. By 2030, climate change is projected to delay China's rapid progress towards reducing WSH-attributable infectious disease burden by 8–85 months.

This development delay summarizes the adverse impact of climate change on WSH-attributable infectious diseases in China, and can be used in other settings where a significant health burden may accompany future changes in climate even as the total burden of disease falls owing to non-climate reasons. Furthermore, this present analysis indicates the need to acknowledge the net delay in development that climate change driven by greenhouse gas emissions will impose. The impact of climate change on WSH-attributable disease most likely extends beyond the burden of disease discussed by Hodges *et al.* , for the study focuses on the impacts associated with climate change and does not incorporate the development delays imposed by climate change due to greenhouse gas emissions. Therefore, it is probable that the true development delay imposed by climate change on China is much higher than stated in this study.

Drought Impacts on Children's Respiratory Health in the Brazilian Amazon

Drought conditions in Amazonia are associated with increased fire incidence, enhancing aerosol emissions with degradation in air quality. On average, the Brazilian Amazon experiences extreme flood or drought once every ten years (Smith *et al.* 2014). However, in 2005 and 2010, only a five-year period, two mega droughts have occurred in the Amazon. Although the 2005 drought was the first in 100 years, the second drought occurred only five years later in 2010. Environmental impacts of drought include tree mortality from water deficits and social impacts include lack of food, lack of medical supplies, isolation of communities, and even health problems. Health issues arise because during droughts, wind erosion in deforested areas causes soil particles and microbes to be blow into the air, creating and exacerbating respiratory problems and triggering allergies. Furthermore, droughts have a positive correlation with fire incidences – in Amazonia, droughts can lead to over 30% increase in fire occurrence. This too leads to more hazardous health issues as smoke from fires tends to carry fine Particulate Matter particles ($PM_{2.5}$), that when inhaled, may reach deep into the lungs, causing irritation of the throat, lungs, and eyes. The primary location for fires within the Amazon is centered around the southern and eastern periphery where 85% of fires occur, emitting as much as 300-600 mg/m^3 of PM_{10}

per 24 hours and up to 400 mg/m³ of $PM_{2.5}$ per 24 hours during the dry season (Smith *et al.*). Measurements carried out in southern Amazonia demonstrated that exposure to $PM_{2.5}$ have positive associations with children's respiratory health. This increase of 10 mg/m³ $PM_{2.5}$ has shown simultaneous correlation with a 5.6% and 2.9% increase in outpatients in Rio Branco, Acre State, and Alta Floresta, Mato Grosso State, respectively.

Despite recent demonstrations of the impact of droughts and fire on tropical ecosystems, there is still a lack of large-scale assessments that inhibit testing of how these droughts would affect the tropical population's health. Therefore, the availability of operational satellite-derived rainfall from the Tropical Rainfall Measuring Mission satellite (TRMM); active fires and aerosol from Terra/MODIS satellite; and deforestation rates from the National Institute for Space Research (INPE) datasets; together with the geo-spatial information about socio-economic indices from the Brazilian Institute of Geography and Statistics (IBGE) and hospitalizations from the Brazilian Health System (SUS), provide a unique opportunity to quantify the sensitivity of children's respiratory health to environmental changes induced by recent droughts in the whole Brazilian Amazon (Smith *et al.*). By determining the relationships between respiratory health and recent Amazonian drought events, an approximation of the expected responses of health to future climate conditions can be provided. Thus, Smith *et al.* tested whether the incidence of respiratory diseases during two major droughts (2005 and 2010) was statistically dependent on drought-related environmental changes and socio-economic factors by using Geographically Weighted Poisson Regression models (GWPR) for the entire Brazilian Amazon.

In order to establish the extent and duration of the drought conditions, Smith *et al.* calculated rainfall, active fire, and aerosol anomalies as a departure from the 2001–2010 long-term mean. Data demonstrated that peak hospitalizations in Acre generally occur at the end of the wet season, however during the 2005 drought, this peak was displaced and in accordance with the fire season, indicating a direct effect. High concentrations of smoke were recorded in Rio Branco during September 2005, which could explain the local parameter estimate for aerosol in this region. (Human Development Index) HDI was significantly positive in drought-affected municipalities in Rondônia, western Mato Grosso, Pará, and Maranhão States, but not around the epicentre of the 2005 drought in Acre (Smith *et al.*).

Additionally, despite the 2010 drought being of greater magnitude than the 2005 drought, the impact on respiratory diseases was not as severe. The strength of HDI in 2010 in relation to the environmental variables may be due to the large spatial extent of the drought, which disguised localized environmental impacts, compared to 2005 when the drought was concentrated around Acre State. Therefore, drought events deteriorated children's respiratory health particularly during 2005 when the drought was more geographically concentrated. Aerosol was significantly and positively related to respiratory disease incidence in southern Pará. However, along the eastern and southern edge of the region, aerosol loads do not increase the number of hospitalizations. This may be because the population living in these locations are accustomed to air pollution exposure as this area sees most fires occurring year on year. Furthermore, the 2010 drought experienced fewer fires and anomalous aerosol loads compared to 2005.

This is the first analysis of the impacts of drought on respiratory health of children under-five years at the scale of the whole Brazilian Amazon. Despite the wide trend of respiratory diseases peaking at the end of the wet season, drought condition exacerbates the incidence of respiratory diseases in children during the dry season. Aerosol, followed by HDI, was the primary driver of hospitalizations in drought-affected municipalities during 2005 (Smith *et al.*). Conversely, during the drought of 2010, HDI overcame the impact of aerosol, probably due to the decrease in aerosol emissions associated with a reduction of 1.9% in fire incidence in 2010 compared to 2005.

This study brings a new dimension into the debate around climate and environmental change impacts in tropical nations. It is now possible to conclude that in the Brazilian Amazon, not only are forests threatened by drought and fires, but also human populations exposed to health-hazardous agents. However, it is encouraging that by efficiently enforcing fire control legislation and curbing fire usage, policy makers could mitigate fire impacts on ecosystems and on human health (Smith *et al.*). Adaptation measurements are also necessary in terms of establishing hospitals in critical areas and planning for greater demand on health services during these crucial drought periods. These policies, together, have huge potential in ensuring better life quality for local populations and possibly improving and minimizing public expenditure for treatment and life costs during future drought events.

Climate Warming will not Decrease Winter Mortality

It has been commonly assumed by policymakers and health professionals that harmful health impacts of anthropogenic climate change will be partially offset by a decline in excess winter deaths (EWDs) in temperature countries as winters warm. However, over the past few decades, the UK and other temperate countries have also simultaneously experienced better housing, improved health care, higher incomes and greater awareness of the risks of cold. Therefore, it is possible that the link between winter temperatures and EWDs is not as direct.

Staddon *et al.* (2014) investigated the key causes that underlie year-to-year variations in EWDs. They demonstrated that three distinct periods in EWD changes were apparent. In 1951–1970, EWDs exhibited very high year-to-year variation and a strongly decreasing overall trend. Then in 1971–2000, year-to-year variation EWDs halved compared with the preceding period and the decreasing trend continued, albeit less strongly. Lastly in 2001–2011, year-to-year variation was very small and the EWD rate was flat. Furthermore, by analyzing more recent data and carrying out rolling correlation analysis on time-detrended data, Staddon *et al.* show unequivocally that the relationship of year-to-year variation in EWDs with the number of cold days in winter of less than 5 °C that was evident until the mid 1970s has disappeared, leaving only the incidence of influenza-like illnesses to explain any of the year-to-year variation in EWDs in the past decade. Though EWDs continue to exist, winter cold severity is no longer a good prediction of the numbers affected.

Staddon *et al.* used a threshold model to identify a strong relationship between the annual number of cold days and EWDs before the mid-1970s and to show that this relationship has since disappeared. The relationship between mortality and local daily temperature is variable and specific to local areas and it is likely that the exposure–response relationship of daily mortality to temperature will have changed over the past decades in response to improved housing, health, and wealth. This data analysis concludes that there is no evidence that EWDs in England and Wales will fall if winters warm with climate change. These findings have important implications for climate change health adaptation policies and for health policies in general.

Potential increases in future winter temperature volatility suggests that EWDs are more likely to rise than fall. Additionally, regardless of whether climate-change-induced winter temperature volatility increases the risk of EWDs, the absolute number of EWDs may increase in the future because of increases in the population. Thus, it is important that the recent policy focus on protecting the population from heatwaves should not be at the expense of preventing the much more numerous EWDs. Energy efficiency regulations and government retrofitting initiatives to improve the thermal efficiency of older homes should continue to benefit both health and climate change mitigation. Staddon *et al.* also believe that particular attention should also be paid to public health initiatives to reduce the risk of infection with flu-like illnesses, including promoting the influenza vaccination, and urgently reducing greenhouse gas emissions to mitigate against climate and weather change also essential.

Global Diets Link Environmental Sustainability and Human Health

Due to rising incomes and urbanization, traditional diets are being replaced by diets that are higher in refined sugars, refined fats, oils, and meats. By 2050, if these dietary trends are unchecked, they can become a major contributor to an estimated 80% increase in global agricultural greenhouse gas emissions from food production and global land clearing, which can result in species extinction (Tilman *et al.* 2014). Additionally, these dietary shifts are greatly increasing the incidence of type II diabetes, coronary heart disease, and other chronic non-communicable diseases that lower global life expectancies. Because the global dietary transition directly links and negatively affects human and environmental health, it is one of the greatest challenges facing humanity. If alternative diets that offer substantial health benefits are widely adopted, there is potential to reduce global agricultural greenhouse gas emissions, reduce land clearing and resultant species extinctions, and even help prevent such diet-related chronic non-communicable diseases. Therefore, the implementation of dietary solutions to the tightly linked diet-environment-health trilemma, although a global challenge, is an important and valuable opportunity to improve the environment and human health. Impactful solutions will not be easily achieved and will require analyses of the quantitative linkages between diets, the environment, and human health. Tilman *et al.* focus their study on these solutions, along with the efforts of nutritionists, agriculturists, public health professionals, educators, policy makers, and food industries.

In this study, Tilman *et al.* compile and analyze global-level data in order to quantify relationships between diet, environmental sustainability, and human health, to evaluate potential environmental impacts of the global dietary transition, and to explore prospective solutions to the diet-environment-health trilemma. Their methods to do so include expanding on earlier food lifecycle analyses (LCAs) by compiling all published LCAs of GHG emissions of food crop, livestock, fishery, and aquaculture production systems. They then use 50 years of data for 100 of the world's most populous nations to analyze global dietary trends and their motivators and use this information to forecast future diets assuming these past trends continue.

Although diets differ within and among nations and regions for a variety of climatic, cultural and historic reasons, diets have been changing in fairly consistent ways as incomes and urbanization have increased globally during the past five decades. Data obtained by Tilman *et al.* indicate several trends amongst global diets. As annual incomes (per capita real gross domestic product, GDP) increased from 1961 to 2009, total protein per capita demand increased 750% in the 15 richest nations relative to the 24 poorest nations, but legume protein demand decreased as animal protein demand increased. A second trend within and among economic groups is the income-dependent increase in demand for calories from refined fats, refined sugars, alcohols and oils. The last trend is that total per capita caloric demand also increased with income. These three trends show that global dietary changes and associated with increased income, which in itself is associated with urbanization and industrial food production. Combining these trends with forecasts of per capita income for the future decades, Tilman *et al.* estimate that relative to the average global diet of 2009, the 2050 global-average per capita income-dependent diet would have 15% more total calories and 11% more total protein, with dietary composition shifting to having 61% more empty calories, 18% fewer servings of fruits and vegetables, 2.7% less plant protein, 23% more pork and poultry, 31% more ruminant meat, 58% more dairy and egg and 82% more fish and seafood.

Diet has a significant impact on human health. Many of the world's poorest people have inadequate diets, and would have improved health were their diets to include more essential fatty acids, minerals, vitamins and protein from fish and meats and added calories and protein from other nutritionally appropriate sources. In contrast, diets of many people with moderate and higher incomes are shifting towards diets that are high in processed foods; refined sugars, refined fats, oils, and meats have contributed to 2.1 billion people becoming overweight or obese. These dietary shifts and resulting increases in body mass indices (BMI) are further associated with increased global incidences of chronic noncommunicable diseases, especially type II diabetes, coronary heart disease, some cancers, and with higher all-cause mortality rates. These diseases are predicted to become two-thirds of the global burden of disease if these dietary trends persist.

In order to quantify the effects of alternative diets on mortality and these chronic noncommunicable diseases, Tilamn *et al.* compiled and summarized results of studies encompassing ten million person-years of observations on diet and health. The results demonstrate that relative to conventional omnivorous diets, across the three alternative of Mediterranean, pescetarian, and vegetarian diets, incidence rates of type II diabetes were reduced by 16%–41% and of cancer by 7%–13%, while relative mortality rates from coronary heart disease were 20%–26% lower and overall mortality rates for all causes combined were 0%–18% lower. This summary illustrates the magnitudes of the health benefits associated with some widely adopted alternative diets. Furthermore, though these patterns do not necessarily indicate that healthier diets are necessarily more environmentally beneficial, not that more environmentally beneficial diets are healthier, they do, however, demonstrate that there are alternative dietary options that should substantially improve both human and environmental health.

Dietary composition also strongly influences greenhouse gases (GHGs) as global agriculture and food production release over 25% of all GHGs. GHG emissions were found to vary widely among foods – relative to animal-based foods, plant-based foods have lower GHG emissions. Diet-driven increases in global food demand and increases in population are also leading to pollution of fresh and marine waters with agrochemicals and clearing of tropical forests, savannas, and grasslands – threatening species with extinction. Using LCA emission data, Tilman *et al.* calculated annual per capita GHG emissions from food production ('cradle to farm gate') for the 2009 global-average diet, for the global-average income-dependent diet projected for 2050, and for Mediterranean, pescetarian and vegetarian diets. Global-average per capita dietary GHG emissions from crop and livestock production would increase 32% from 2009 to 2050 if global diets changed in the income-dependent ways. All three alternative diets could reduce emissions from food production below those of the projected 2050 income-dependent diet, with per capita reductions being 30%, 45% and 55% for the Mediterranean, pescetarian and vegetarian diets, respectively. However, minimizing environmental impacts does not necessarily maximize human health. For example, prepared items high in sugars, fats, or carbohydrates can have low GHG emissions but be less healthy than foods they displace. Solutions to the diet–environment–health trilemma should seek healthier diets that have low GHG emissions rather than diets that might minimize only GHG emissions.

Changes towards healthier diets can also have globally significant GHG benefits. From 2009 to 2050, the global population is projected to increase by 36%. When combined with the projected 32% increase in per capita emissions from income-dependent global dietary shifts, the net effect is an estimated 80% increase in global GHG emissions from food production. In contrast, there would be no net increase in food production emissions if by 2050 the global diet had become the average of the Mediterranean, pescetarian, and vegetarian diets.

The analyses by Tilman *et al.* demonstrate that there are plausible solutions to the diet–environment–health trilemma, diets already chosen by many people that, if widely adopted, would offer global environmental and public health benefits. Unless the nutrition transition that is under way is changed, diabetes, chronic heart disease and other diet-related chronic non-communicable diseases will become the dominant global disease burden, often affecting even the poorer members of poorer nations for whom appropriate health care is unavailable. Furthermore, the dietary choices that individuals make are influenced by culture, nutritional knowledge, price, availability, taste and convenience, all of which must be considered if the dietary transition that is taking place is to be counteracted. The evaluation and implementation of dietary solutions to the tightly linked diet–environment–health trilemma is a global challenge, and opportunity, of great environmental and public health importance.

Solar Ultraviolet Radiation in Changing Climate

In the early 1970s, chlorofluorocarbons that were widely used as refrigerants and propellants were believed to reach the stratosphere and catalyze the destruction of ozone molecules there. Later in 1985, evidence of an "ozone hole" over Antarctica was first published, and progressing images have been symbolic of human influences on the global environment. Fortunately, large-scale depletion of stratospheric ozone and high levels of ultraviolet radiation have been relatively avoided by the success of the 1987 Montreal Protocol on Substances that Deplete the Ozone Layer. Although solar radiation is essential to life on Earth, its ultraviolet component may damage both living organisms and non-living matter. Ultraviolet radiation is divided into three wavelength bands: UV-A (315–400 nm), UV-B (280–315 nm), and UV-C (100–280 nm). UV-C radiation is the most damaging but is completely filtered out by the Earth's atmosphere and does not reach the surface. UV-A, however, passes rather unperturbed through the ozone layer and thus is the largest component of ground-level solar radiation. The effects of UV-A are wide ranging, and have been shown to influence air quality, aquatic and soil processes, and human health. These data helped to establish the impetus behind the Montreal Protocol. Implementation of the Montreal Protocol has drastically curtailed production of chlorofluorocarbons and other ozone depleting substances (ODS).

The unequivocal warming of the climate may be involved in stratospheric ozone depletion independently of concentrations of ODS in the atmosphere. Specifically, global warming may affect stratospheric ozone depletion by increasing its water content, which can act to accelerate ozone destruction by activating gas phase hydrogen oxides, enhancing the catalytic cycle, and by increasing the surface area of stratospheric aerosol particles on which ozone-depleting halogen molecules can be activated.

Solar UV radiation has a profound impact on atmospheric chemical homeostasis. Solar UV radiation is involved in the balance between the production of chemicals that help to clean the atmosphere as well as those that generate photochemical smog. This balancing act is a complex one and is itself influenced by many factors such as precipitation, vegetation cover, and agricultural intensification. In fact, research investigating the relationship between solar UV-B radiation and agriculture has generated positive results that have been exploited in horticulture to improve food production and quality. However, long term effects between UV-B radiation and other environmental stress factors are still unknown.

The extent and duration of periods of ice and snow cover on oceanic and inland waters has been decreasing steadily over the last few decades, changing the exposure patterns and intensity of solar UV radiation on aquatic ecosystems ecosystems. This shift in UV exposure has the potential to alter the fundamental structure of the living organisms within aquatic ecosystems, negatively impacting the ability of phytoplankton to carry out photosynthesis and therefore disrupting the entire metabolic chain.

Extensive consideration has been given to the effect of UV radiation of human health and disease. Evidence currently shows that UV radiation can have detrimental effects on human health, causing cataracts as well as many different types of skin cancers. In fact, it is believed that increased durations of solar UV exposure has led to an increase in the prevalence of both non-melanoma skin cancers (NMSCs) and cutaneous melanomas (CMs) within fair skinned populations. Modeling studies have estimated that the prevalence of these cancers would be far greater than they currently are had it not been for the Montreal Protocol and its amendments. However, it has also been noted that sunlight is involved in vitamin D synthesis. Therefore, the benefit gained by acute exposure to sunlight UV radiation in small quantities must be balanced with the risk posed by long durations of intense exposure leading to sunburns and predisposing to cancer. Ultimately new research is needed to address the breadth and complexity behind the effects of global climate change and sunlight UV radiation on the ecosystems and human populations of this planet.

New insights and knowledge generated by the research response to the Montreal Protocol reveals that complete understanding of the UV-ozone-climate links still has some holes. New research is necessary to uncover the potential risks and benefits across the atmosphere and biosphere as a result of the coupled ozone depletion-climate change interactions.

Conclusions

Changes in infectious disease transmission patterns are a likely major consequence of climate change. Therefore, it is essential to learn more about the underlying complex relationships, using more complete and validated models, and applying this information to the prediction of future impacts. The individual papers discussed in this chapter help to explain relationships between climate variability and various factors of human health including transmission of several infectious diseases, human diet, and solar ultraviolet radiation. These findings may be beneficial in helping public health officials predict epidemics and prepare for the effects of climate change on infectious diseases through the implementation of preventative public health measures.

References Cited

Hodges, M., Belle, J. H., Carlton, E. J., Liang, S., Li, H., Luo, W., Freeman, M. C., Liu, Y., Gao, Y., Hess, J. J., Remais, J. V., 2014. Delays in reducing waterborne and water-related infectious diseases in China under climate change. Nature climate change 4, 1109–1115.

Onozuka, D., 2014. Effect of non-stationary climate on infectious gastroenteritis transmission in Japan. Scientific reports 4, 1-6.

Onozuka, D., Chaves, L. F. 2014. Climate Variability and Nonstationary Dynamics of Mycoplasma pneumoniae Pneumonia in Japan. PloS one 9, 95447.

Onozuka, D., Hagihara, A., 2014. The association of extreme temperatures and the incidence of tuberculosis in Japan. International journal of biometeorology 1-8.

Smith, L. T., Aragão, L. E., Sabel, C. E., & Nakaya, T., 2014. Drought impacts on children's respiratory health in the Brazilian Amazon. Scientific reports 4, 1-8.

Staddon, P. L., Montgomery, H. E., Depledge, M. H., 2014. Climate warming will not decrease winter mortality. Nature Climate Change 4, 190-194.

Tilman, D., Clark, M., 2014. Global diets link environmental sustainability and human health. Nature 515, 518-522.

Williamson, C. E., Zepp, R. G., Lucas, R. M., Madronich, S., Austin, A. T., Ballaré, C. L., Norval, M. *et al.* 2014. Solar ultraviolet radiation in a changing climate. Nature Climate Change 4, 434-441.

River Delta Systems and Climate Change

Rebecca Herrera

A river delta is the name for the geological landform that forms where a river ends and flows into an ocean, sea, estuary, lake, reservoir, or valley. Deltas are named for their triangular shape, caused by slow speeds at flat elevations. River deltas are sensitive ecosystems that play host to high biodiversity levels. Human interference of river and delta systems through levies, dams, and other obstructions can and have had detrimental and long lasting effects on delta ecosystems.

People ought to value deltas purely for their natural presence, much like someone who has never seen the Grand Canyon values the fact that there is one, a phenomena called existence value. But river deltas also perform a number of priceless ecosystem services. For example, millions of Californians get their water supply from the San Joaquin River delta, Batswana who live in the inland Okavango delta thrive on a diet of fish, and the ancient Mesopotamians flourished in the aptly named Fertile Crescent of the Tigris and Euphrates Rivers. Today, river deltas still provide rich soils that support robust agricultural sectors in addition to supplying sand and gravel for construction and public work projects. However, human interference in delta dynamics has repercussions for the maintenance of a healthful ecosystem. The Mississippi River delta has lost thousands of square kilometers to open water due to rising sea levels while the Ebro River basin of North-East Spain is experiencing fluxes in toxicology caused by industry and agriculture. These anthropogenic factors contributing to the decline of delta health should be mitigated. Thankfully, numerous individuals are invested in recuperating the world's deltas. Through gaining a better understanding of delta systems and functions, scientists gain valuable insight into how humans can simultaneously protect and make use of the world's river delta systems.

Modeling the effects of Urbanization on local climates in the Pearl River Delta

It has become clear that cities and other urban localities experience warmer temperatures than their rural counterparts. The urban heat islands of megatropolises can have more extreme effects on a specific area than global temperature rises have on the same area. This phenomenon was studied by Wang *et al.* in the Pearl River Delta region of Southern China in the coastal zone of the Guangdong Province, an area that has experienced dramatic economic development over the last thirty years. The authors attempt to gain a more thorough and comprehensive understanding of the effects of urbanization on regional and local climactic indicators over an extended period of time. The results showed an overall average temperature increase, an decrease in daytime temperature ranges, a decrease in near surface water vapor quantities, a decrease in the annual number of precipitation days, an increase in annual precipitation, and a decrease in average wind speeds across urbanized zones.

The Pearl River Delta of Southern China has experienced remarkable economic growth and urbanization over the past thirty years; urban areas account for over 60% of land use in the region, making it an apt and appropriate place to study the effects of urbanization on local climates. The need to study the consequences of urbanization on local and regional climate is important as more of the world's population migrates to urban areas and as more urban areas develop. The authors studied urbanization and climate indicators through a model called the Advanced Research core of the Weather Research and Forecasting system that conducted a pair of simulations using two different representations of urbanization on the Pearl River

Delta region. The efficacy of the model was ensured against results of previous models along with prior data from the year in question.

The results showed an increase in average temperatures in all seasons while there was a decrease in the temperature range. Changes in the surface albedo through the shading effect likely increased solar radiation absorption while tall buildings likely trapped such radiative heat in synthetic 'street canyons.' The temperature range decreased because the average minimum temperature rose more than the average maximum temperature. 'Surface roughness' also increased, according to the authors. Because of the highly mixed distribution of buildings in urban areas as well as the high concentration of buildings in urban areas, friction and drag on winds increases and average wind speeds fell in urban areas. Surface water vapor concentrations in urban areas also fell as urban areas have a high area of impervious surfaces, which decrease the amount of water available for evaporation. Finally, while the model showed a decrease in the number of days with precipitation, it did show an increase in overall precipitation, indicating an increase in localized 'extreme' weather events.

It is evident that urbanization does in fact have a significant effect on local and regional climate in the Pearl River Delta. It is also likely that these findings will hold true for other urban areas across the world. It is necessary to continue studying the effects of urbanization on the planet as more individuals migrate to cities. Additional parameters to consider for future studies include the effect urbanization and urban head has on airborne pollutants.

Sediment Eco-toxicological Status for the Evaluation of Surface Water Quality: the Ebro River Basin

Roig *et al.* (2014) assess various sites along the Ebro river basin and collect information on each site's eco-toxicological status. The data are then used to complement traditional means of evaluating surface water quality. The researchers compare the effectiveness and viability of different eco-toxicity tests, otherwise known as bioassays, performed with freshwater sediments, and evaluated the relationship between ecological status, pollutant concentrations and pore water and sediment eco-toxicity to make recommendations to the European Water Framework Directive (WFD). Their findings showed a high correspondence of the eco-toxicological measurements to prior measurements of ecological status, especially when ecosystem disruption due to numerous stressors was observed.

The Ebro River basin, located in North-East Spain, is characterized by the interannual variability associated with its Mediterranean climate. The researchers measured data at 12 sampling sites and one reference site. Sampling sites were chosen to create a representative picture of the river basin, while sampling adjacent to important agricultural, industrial, and urban areas. The reference point experiences relatively low environmental and anthropogenic stressors in comparison to the other sites.

Basic measurements of humidity, porosity, percentage fine and organic matter, organic carbon, ammonium, and pH were taken at all 13 sites. To determine if potentially toxic metals associated with sulfides were present in sediments, researchers analyzed acid-volatile sulfide (AVS) and simultaneously extracted metals (SEM), and then checked to see which was greater. If SEM is greater than AVS, it is likely the sediment is nontoxic. Researchers also measured concentrations of potentially toxic elements, including arsenic, cadmium, chromium, copper, mercury, nickel, lead, and zinc. The ecotoxicity of pore water was evaluated using the bacterium $PM_{2.5}$, a freshwater green algae *Pseudokirschneriella subcapitata*, and the crustacean *Daphnia magna*. Whole sediment ecotoxicity was evaluated again with *V. fischeri* and *D. magna*, in addition to the benthic diatom *Nitzschia paleax*, and the midge *Chironommus riparius*.

All study sites showed poor chemical status according to the WFD's strict guidelines, however, some sites were also considered of good or high quality when evaluated using a biological, physicochemical, and hydro-morphological integrated approach. High coincidences were found when comparing the bioassays to prior methods, proving a high efficacy for the new eco-toxicological approach.

Measuring surface water toxicity through bioassays is an exciting prospect in that it is both cost-effective and rapid to complete for both pore water and whole sediments. While traditional, chemical monitoring provides a detailed picture, it fails to interpret the effect of toxins and varying levels of toxicity on biota. The eco-toxicity method provides a direct measure of significant toxicity of contaminated environments.

Modeling CO_2 and CH_4 Fluxes in the Arctic using Satellite data

The peatlands and tundras of the Arctic perform vital ecosystem services to the earth through their ability to sequester carbon (CO_2) and methane (CH_4) and function as a carbon sink. The ability of the permafrost in the peatlands and tundra ecosystems of the Arctic to continue to function as a natural reservoir for carbon and methane may be disrupted by rising global temperatures that increase the rate of soil decomposition. Watts *et al.* (2014) integrate a terrestrial carbon flux (TCF) model to include a newly developed CH_4 emissions algorithm. The new TCF model simultaneously assesses CO_2 and CH_4 fluctuations and the corresponding net ecosystem carbon balance (NECB), which is contingent upon gross primary productivity (GPP) subtracted from ecosystem respiration. The integrated TCF model uses data gathered through satellite remote sensors to assess fluxes in CO_2 and CH_4.

The researchers examined data from six pan-Arctic peatland and tundra sites in Finland, Sweden, Russia, Greenland, and Alaska that are representative of wet permafrost ecosystems with varied terrain and vegetation. While previous assessments of CO_2 and CH_4 fluxes have been effective at providing localized measurements, extrapolating the data to larger geographic regions has proved challenging. Efficacy of the integrated model was proven through a comparison of the model to previous data collected through previous NECB measurements.

Although the model estimated CH_4 emissions to be small when compared to ecosystem respiration per annum, the model also showed that these CH_4 emissions reduced study site NECB by 23%. The model indicates a net carbon sink for five of the six monitored sites; meaning ecosystem respiration exceeded annual GPP. Excluding two mitigating study sites, the model estimates annual CO_2 and CH_4 fluxes have significant potential to contribute to global warming.

Reduced NECB has global effects on climate change. CO_2 and CH_4 are greenhouse gasses whose increased concentrations contribute to global surface temperature increases and perpetuate a positive feedback loop of increased GHG concentrations and increased global temperatures as permafrost continues to thaw. Biogeochemical models that monitor CO_2 and CH_4 ought to continue to be improved and studied as Arctic peatlands and tundras continue to loose their permafrost.

Anthropogenic Effects on Greenhouse Gas (CH_4 and N_2O) Emissions in the Guadalete River Estuary (SW Spain)

Burgos *et al.* (2014) discuss seasonal variations of the greenhouse gases methane (CH_4) and nitrous oxide (N_2O) in the Guadalete River Estuary ending in the Cadiz Bay of southwestern Spain. They found that greenhouse gas concentrations were higher in the more inland parts of the estuary compared to the mouth of the river. Concentrations of methane and nitrous oxide varied depending largely upon the seasonal precipitation regime. It was also observed that the Guadalete Estuary acted as a source, rather than a sink, of greenhouse gases throughout the entire year, as observed by measuring the fluxes of CH_4 and N_2O from the Estuary.

Eight sampling sites were chosen along an eighteen kilometer stretch along the Guadalete River Estuary. The first was at the mouth of the River in a town of approximately 89,000 inhabitants, and the eighth was inland adjacent to a city of approximately 215,000 people. The majority of the land adjacent to the Guadalete River and Estuary is used for irrigated crops such as sugar beet, cotton, and wheat. Samples of surface water were taken seasonally over the year 2013.

Seasonal variations of seawater parameters including salinity, temperature, dissolved oxygen, chlorophyll, dissolved organic carbon, dissolved organic nitrogen, particulate organic carbon, and particulate organic nitrogen were found to be significant and all parameters except temperature were statistically different regarding the distance to the river mouth. Additionally, nutrient content increased from the mouth of the river towards the inner estuary in all seasons.

Dissolved greenhouse gas concentrations of methane and nitrous oxide increased from the mouth of the river inland toward the freshwater estuarine zone. This distribution can likely be attributed to organic matter inputs sourced from both the river itself as well as anthropogenic inputs like sewage and crop runoff. It is important to gain a thorough understanding of to what degree nutrient and gas exchange occurs in order to better target policies at reducing and eliminating sources of anthropogenic contributions to greenhouse gas inputs into estuarine ecosystems.

Climate Change and Agricultural Water Resources: A vulnerability assessment of the Black Sea catchment

Agriculture in the Black Sea catchment is a large share of the region's economy and will experience a number of changes thanks to climate change. Bär *et al.* designed a new climate and agriculture assessment combining two popular techniques in order to better assess the vulnerability of the Black Sea catchment's agriculture. By combining the DPSIR and the vulnerability concept, Bär *et al.* were able to see that rising temperatures, decreased precipitation, or both created different outcomes across the catchment. While plant growth conditions mostly improve across the Black Sea catchment with rising temperatures, decreased precipitation and decreased irrigation potential caused deteriorating agricultural conditions.

By constructing a combination of the Driver-Pressure-State-Impact-Response (DPSIR) model and a vulnerability assessment, the researchers create a more suitable and operationalized model framework. DPSIR shows the causal relationships between elements, as drivers create pressures that change states and create impacts that warrant responses. Researches identify one of the parameters of DISPR and follow it through the chain of drivers, pressures, states, impacts, and responses to ultimately assess how a driver affects the rest of the ecosystem and how to best mitigate said driver. Vulnerability assessments on the other hand show what areas are susceptible, or unable to cope with, adverse effects of climate change, including climate variability and extremes.

Interdisciplinary Assessment of Sea-level Rise and Climate Change Impacts on the lower Nile delta, Egypt

The Mediterranean region will continue to experience climate change in much of the same way as other arid regions around the world. Susnik *et al.* (2014) set our to find our how the Nile River delta in Egypt experiences more intense droughts and water shortages, rising regional temperatures, an increased frequency of flash flood events, and sea level rise. It is critical to understand of how these climactic changes will impact the people residing in the delta. Susnik *et al.* take an integrated and interdisciplinary approach to studying the effects of sea level rise (SLR) on the lower Nile delta and the greater Alexandria areas by analyzing the results of three complementary projects; which reveal that water overexploitation exacerbates land subsidence and accelerates saline intrusion of soils and groundwater which has radiating effects on employment as well as placing additional pressure on agricultural lands and regional development.

The projects are CLIMB (Climate Induced changes on the hydrology of Mediterranean Basins), WASSERMed (Water Availability and Security in Southern Europe and the Mediterranean), and CLICO (CLImate change hydro-COnflicts and human security). These projects focus on the various socio-physical water-related security issues throughout the Mediterranean and comprise the CLImate-induced changes on WAter and SECurity (CLIWASEC) study.

The CLIMB research was based on the implementation of multi-model hydro-meteorological ensembles and the comparison of changing hydrological quantities from the reference period of 1971-2000 to the projection period of 2041–2070. For the Nile Delta, CLIMB found that annual precipitation will fall, there will be moderate shifts in seasonal rainfall patterns, and the area will have an increased susceptibility to drought. The Greater Alexandria region according to the results is highly vulnerable to urbanization, ground water and soil salinization, pollution, land degradation, and poor management practices. Local stakeholders see population growth and urbanization as the main pressures on water resources over the next 20 years.

WASSERMed used "System Dynamics Modeling" to examine water security concerns in the water, food, and economic sectors in the Rosetta region of the Nile Delta. WASSERMed determined that water in the Nile delta is an overexploited resource. As total SLR increases, the amount of agricultural land loss increases to a maximum of 13% at 0.5 m SLR. The outcomes are not all grim however, as in cropping pattern simulations, it was shown that the overall water balance improves and water use drops as farmers replace water intensive rice crops are replaced with cotton, vegetables, and wheat; crops already planted in the area. By doing this, farmers have the potential to increase their crop yield and improve their livelihoods as well as mitigate SLR and climate change impacts on their crops by reducing their water demand.

CLICO studies and predicts new susceptibilities and security concerns evolving from an adaptation scenario to SLR of large-scale resettlement. Because 35% of the densely populated areas of the Greater Alexandria region reside below sea level, some see preventative resettlement as potentially the only viable response to SLR. Through 500 household surveys, the CLICO team found that households have a limited capacity to migrate, especially farmers and fishers whose livelihoods are tied to their surroundings. Up to 20% of respondents reported that they would refuse to voluntarily participate in a government resettlement program. If resettlement fails and SLR continues to climb, people may be forced into informal housing, compromising their livelihoods and increasing the likelihood of conflict.

A more complete understanding of the pressures of climate change and sea level rise in the Nile delta and Greater Alexandria region of Egypt can help policy makers better manage regional development in regards to climate and social change.

Sand as a Stable and Sustainable resource for Nourishing the Mississippi River delta

The Mississippi River delta in the Southern United States has lost approximately 5,000 square kilometers of low lying wetlands over the last 80 years to open water as a consequence of anthropogenic and natural factors. Nittrouer and Viparelli (2014) analyzed sediment load in the lower Mississippi River through a one-dimensional numerical model. Their model predicts that the sand load feeding the delta from further north along the Mississippi River will "decrease only gradually over the next several centuries, with an estimated decline from current values of no more than about 17% within the coming six centuries."

Primarily a result of sea level rise, increased leveeing, and excavation of the delta for navigation channels, the Mississippi delta is loosing land to open water at an alarming rate. Sediment that contributes to the inherent marshy qualities of the delta is important because sediments form stable platforms on which vegetation may settle. Because humans have colonized the Mississippi River delta, the levees and dams contributing to the degraded delta cannot simply be removed. Instead, targeted and responsible engineering practices must be implemented to protect society and infrastructure while effectively imitating natural processes that contribute to the maintenance of a healthy delta ecology. One such example of strategic engineering are 'land-building' diversions in which a structure captures a large amount of sediment from the main channel and delivers the sediment to delta drowning zones.

The one-dimensional numerical model of morphodynamics the authors used mimicked riverbed degradation in response to a reduction in sand supply upstream. The model takes into account hydraulics, and total sand load, as well as sediment continuity. Nittrouer and Viparelli chose to model a 75% decrease in land load upstream as prior studies have shown that sediment load in the Lower Mississippi River basin have decreased by approximately 75% since significant human interfering in the Mississippi River.

Final analysis revealed that sand from the upstream portion of the Mississippi River could provide a stable and sustainable supply of sand to the lower delta for the foreseeable future. Additionally, the researchers conclude that the sand supply to the delta has not been inhibited or reduced by the construction of dams along the river in the 1950s.

Dynamics of Landscape Patterns in an inland river delta of Central Asia based on a Cellular automatic – Markov Model

Luo *et al.* (2014) study the Ili River delta of Central Asia in an attempt to identify and analyze landscape pattern changes from 1978-2007 and their possible driving forces, assess the efficacy of a CA-Markov model, and to apply the model to predict future landscape pattern changes for the next 20 years. The total wetland area, including the marsh and floodplain areas, and sandy land areas have remained relatively steady, while desert grasslands have slightly decreased and shrublands within the delta have slightly increased. The CA-Markov model is somewhat effective in the overall simulation of landscape patterns, but has difficulties anticipating all changes due to human disturbances. The model predicted that the landscape of the Ili River delta would improve under Scenario A (based on data from 1990 to 2007) and would degrade under Scenario B (1990–1998).

The study area of the Ili River delta was selected because landscape pattern dynamics of the region are driven by physical environmental factors, such as climate change and runoff flow while direct human impacts have been relatively weak within the study period. The total delta exceeds an area of 10,000 square km of which approximately 37% are wetlands.

The model used and evaluated is a hybrid of cellular automata (CA) and Markov models. In a CA model, the decisions made for each pixel explicitly takes into consideration the state of neighboring pixels. CA models have an explicitly spatially component. A Markov process is a "discrete random process whose future probability depends only on its probability of the previous state" (Luo *et al.* 2014).

Luo *et al.* used four types of data sets to incorporate into the CA-Markov model. Landsat satellite data from the USGS was used to map the land cover distribution, land cover survey data was collected in May 2009 and June 2012 as well as Google Earth™ images were used for assessing the accuracy of the model, hydrological data on fluvial runoff from the Ili River from 1953–1992 and water levels of Lake Balkhash (1953–2005) were used to account for landscape dynamics of wetlands, and finally, socioeconomic data were obtained between 1970 and 2007 along with ancillary topographic data that were used to develop criteria for making decisions for various land cover type transitions in the simulation.

The model was highly accurate in land cover classification. Wetlands, shrublands, and desert grasslands were found to dominate the delta across the study period and total wetland area remained approximately the same from 1978 to 2007. Desert grasslands expanded until 1990 by 20%, but then shrunk back to their 1978 size by 2007, being replaced by shrubland, marshes, and floodplains. Sandylands remained the same overall, but expanded and contracted during the study period.

The Markov Model can only explain changes in the landscape from 1990 to 2007. Luo *et al.* believe the model experienced difficulties predicting earlier landscape changes because of human disturbances. While the CA-Markov model is fine for predicting rivers, desert grasslands, and sandyland areas, its failure to predict changes in lakes and marshes lowers its overall efficacy.

Landscape patterns for 2020 are predicted to remain generally similar to the landscapes of 1998 and 2007. The landscape is expected to improve under Scenario A, and degrade slightly under Scenario B. Possible forces of landscape change are regional climate variability and change and indirect human activities and interferences.

This study is important as the research will provide an additional resource for analyzing underlying causes of landscape pattern dynamics and its results may have far reaching environmental and socioeconomic policy implications for sustainable land use.

Conclusions

Delta ecosystems are fragile yet incredibly useful, and therefore ought to be protected, respected, and used with caution. The more we know about the complexities of river deltas, the better scientists, policy makers, and other stakeholders will be better able to utilize bountiful delta environments while simultaneously protecting these pockets of high biodiversity.

References Cited

Bär, R., Rouholahnejad, E., Rahman, K., Abbaspour, K.C., *et al.* 2014. Climate change and agricultural water resources: a vulnerability assessment of the Black Sea catchment. Environmental Science and Policy, 46, 57-69.

Burgos, M., Sierra, A., Ortega, T., Jorja, J.M. 2014. Anthropogenic effects on greenhouse gas (CH4 and N2O) emissions in the Guadalete River Estuary (SW Spain).. Science of the Total Environment, 503-504, 179-189.

Luo, G., Amuti, T., Zhu, L., Mambetov, B., Maisupova, B., Zhang, C. 2014. Dynamics of landscape patterns in an inland river delta of Central Asia based on a cellular automata-Markov model. Regional Environmental Change, 15, 277-289.

Nittrouer, J., Viparelli, E. 2014. Sand as a stable and sustainable resource for nourishing the Mississippi River delta. Nature Geoscience, 7, 350-354.

Roig, N., Sierra, J., Moreno-Garrido, I., Hampel, M., Gallego, E., Schuhmacher, M., Blasco, J. 2014. Assessment of sediment ecotoxicological status as a complementary tool for the evaluation of surface water quality: the Ebro river basin case study. Science of the Total Environment, 503-504, 269-278.

Susnik, J., Vamvakeridou-Lyroudia, L., Baumert, N., Kloos, J., *et al.* 2014. Interdisciplinary assessment of sea-level rise and climate change impacts on the lower Nile delta, Egypt. Science of the Total Environment, 503-504, 279-288.

Wang, X., Liao, J., Zhang, J., Shen, C., *et al.* 2014. A Numerical Study of Regional Climate Change Induced by Urban Expansion in the Pearl River Delta, China. Journal of Applied Meteorology and Climatology, 53, 346-362.

Watts, J. D., Kimball, J. S., Parmentier, F. J. W., Sachs, T. *et al.* 2014. A satellite data driven biophysical modeling approach for estimating northern peatland and tundra CO_2 and CH4 fluxes. Biogeosciences, 11, 1961-1980.

Natural Resources and Governance Policy: A Regional Survey

Lazaros M. K. Chalkias

In light of climate change and the pressing need for sustainable development paths in Africa, Central and South America, and Asia, natural resources management research is becoming increasingly important in informing national and international policy. As 2015 is the set expiration year for the Millennium Development Goals, the world needs to define a new agenda looking into sustainability from an environmental, economic and social perspective. One avenue to be explored on a national level involves targeted policy, public-private partnerships, and community work. The literature reviewed in this chapter surveys policy changes—and research substantiating such changes or recommendations—from a regional perspective. Special attention is paid to water management and governance.

Evidence from water management systems developing in Namibia and Sweden, indicates solutions common across continents despite the absence of direct interactions between the two countries. Bollig and Schwieger (2014) reported the formation of Water Point Associations on a community basis in Namibia as a successful in both short- and long-term management. In Costa Rica, the highly fragmented network of water governance was found to be functioning most efficiently under guidance of a local leader (Kuzdas *et al.* 2015). However, collaborative work spanning communities to national conservation efforts can lead to amazing results in natural resources modeling and management, as with the case of water management in the Netherlands (De Lange *et al.* 2014).

Similarly, Stoett and Temby (2015) report success in the organizations studies under the framework of transnational networks across North America.

Implementation of successful management and governance of water and other resources will be a challenging, time-, and funds-consuming task. Contrasting evidence to the success of community-based management does not allow for understudied conclusions (Pfaff *et al.* 2014). The climate circumstances, however, and decisions of the international community have expanded the scope of research and started an international and intergovernmental search for more efficient and effective governance regimes.

Water and Game Management in North-Western Namibia

The foundation of Community-Based Natural Resource Management (CBNRM) lies in the community agreement of maintaining the local resources available in an equitable manner and with sustainability in mind. The Namibian Government had transferred the management of natural resources to local governments early on, with some significant advances over the last two decades, in an attempt to ensure sustainability and empower local communities to make decisions and benefit directly from their resources. Bollig and Schwieger (2014) focus on the development of game conservancies and Water Point Associations (WPAs), mainly in North-Western Namibia, where historical and social factors often have created competition and conflict on a local, small scale basis.

Following governmental reforms in the early 1990s, rural communities could form into game conservatories from 1996 and into WPAs from 1997 through the Directorate of Rural Water Supply (DRWS). The conservancy program gained high interest from locals, who explored their social-ecological relations through spatial exploration of the conservancies, zoning for easier land use control and monitoring of wildlife. At times, the authors note involvement from the tourist industry and NGOs for more expansive local management (contracts managed by elected committee). The result is a significantly improved meat distribution and game management process; unfortunately cash flows still lack in transparency or accountability. In WPAs, the principles of equity and fairness guide the management of water as performed by peer evaluation and transparent management.

CBNRM examples can be evaluated based on their effect on the advancement of institutional development in rural areas, the success of decentralization of the resource management, local empowerment, or income changes. Specifically, groups managing Common Pool Resources (CPR) can be evaluated based on their size, monitoring and sanctioning mechanisms, and their perception of fairness and trust. In North-Western Namibia, administrative control of resources has moved away from the government and from the neighborhood level to a community one, accountable to NGOs and central administration, but mostly to the people served. Trust and fairness into the system has been consistently building with hopes that it will improve post-2015.

Apart from the immediate resources-oriented benefit from this degree of organization, Bollig and Schwieger discovered that conservancies, especially, offer a higher degree of safety from outsiders' intrusion and resource exploitation. The fact that group size did not appear to highly correlate with success allows for great potential of further local organization and legislation of such groups that already manage their funding and income well. The next step would allow for the establishment of systematic monitoring and sanctioning system for advancement of their accountability and sustainability of approaches.

Governance Effectiveness Analysis of Socio-Ecological Networks: a Case Study of Stockholm Wetlands

As the effects of climate change become increasingly severe, balancing the demands of ecosystem services with efforts to revert their degradation will call for new institutions and policies. Kininmonth *et al.* (2015) attempted to understand the governance structures and relationships of people that determine the resilience of social-ecological systems. Using motifs (simple network substructures defined to facilitate studying) they examined various system governance structures as influenced by the common-pool resource theory, and the principle of social and ecological alignment. The theoretical models were then applied to 25 municipalities and 408 wetlands in central Stockholm County, Sweden. They found that adaptive governance structures that can manage "uncertainty and surprise" are needed yet many suggested ideas are generally not operational. The conclusion was, therefore, satisfactory with regards to the existing model but it remains not expansive enough to cover all cases.

Applying their framework to wetland management structures they explored the interaction of people as social nodes and resources as ecological nodes. The primary motif studied was composed of three social and two ecological nodes. The model contained only four motif types in which two resource-managers share a resource and showed that the best scenario requires the collaboration of the managers, often with a coordinating actor. A second model proposed with a higher degree of complexity, allowed for consideration of "shared benefits and burdens" to account for effects of resource mismanagement.

The researchers utilized amphibians as model species to derive conclusions about the socio-ecological assessment of wetland heath management. The information was quantified by comparison of the deviations of motif occurrence in the system studied to a random theoretical one, and principal component analysis specified the similarities of municipalities involved in the wetlands management. The results indicated that municipalities in Stockholm utilize coordinating actors and collaboration methods more frequently than expected, probably as a careful response to local-scale wetland management needs. Further analysis indicated very high performance on an "ecological connectivity" approach, yet no causation can be shown with a local government decision to intentionally increase governance effectiveness. Unfortunately, the current model could not account for some of the more complex motifs observed, yet the significance of this study lies in the insight provided by the systematic inclusion of third social actor, a coordinating component.

Participation in Water Governance Regimes: A Case for Rural Costa Rica

Recognizing the vulnerability of people and water systems in rural Central American dry tropics, Kuzdas *et al.* (2015) examined local water governance designed to meet water-related development in those regions. The focus of their work was governance, defined as when political and decision-making powers are shared between groups–instead of the traditional top-down management. In a setting like that, governmental control and effectiveness is limited in favor of civil society and private actors, and governance describes the "collective actions" driving water systems toward the "shared goals" of all sectors. The Guanacaste Province in Northwest Costa Rica was the targeted case study and the research questions asked pertained to: (i) the structure of institutions and actors network, (ii) the extent of individuals' influence to water governance, and (iii) the extent to which "collective knowledge gaps" affect their actions. Using a systems framework to study water supply, delivery, use and outflows, the researchers describe a largely fragmented and de-concentrated water governance system with potential to improve in a polycentric framework.

The City of Nicoya was selected as the primary site of study because its entire water supply comes from Río Potrero via surface flows. The public and private organizations and individuals that comprise the governance network were also deemed more complex when compared to similar regions. Social network analysis was performed to evaluate the activity and cooperativeness of the actors and the results were combined with the evaluation of the physical components of the water management to produce a system and stakeholder map. The results indicated that the National water utility agency (AyA) and the Environment Ministry (MINAET) were largely unsuccessful in coordinated water-related activities despite their mandate. The two agencies did not act coherently while other agencies were absent altogether. Individuals within the organizations and groups on the other hand seemed to perform best as leaders on the local level. With regards to the knowledge gaps identified, the storage capacity of the Potrero-Caimital Aquifer and the characteristics of local groundwater remain unknown and not accessible by the local communities.

Despite the high uncertainty observed around water supply activities by MINAET and AyA, the rural capacity building and other activities of the PC Commission (a grassroots service organization in the area) constitute the domain the most collaborative and active of those studied. Water delivery was found to involve fewer organizations with low cohesion measures and authority centralized to AyA, indicating general governance disorganization. Similar elements determine the outflow domain with the lowest levels of activity and cooperation observed. Water use analysis demonstrated increasing and diverse activity with very limited cooperation among the many actors involved and limited self-evaluation; most of the programs there are controlled by MINAET. The centralized management from the capital was found to create a disconnect with rural branches and despite attempts to create polycentric schemes at a national level, Guanacaste was rarely affected by positive change. The recommendations determined include recognition of the insufficient groundwater availability, the "uneven" governance capacity and potential in local leaders as actors of change.

Comprehensive and Sustainable Water Management in the Netherlands

The Netherlands has been harvesting the benefits of major European rivers (Rhine, Meuse), of accessibility to the sea and high precipitation at the cost of a constant struggle for safety and survival from the elements themselves. These conditions have bestowed great responsibility to the government to plan and prepare for disasters of drought and flood. Following the 1976 drought, the idea of an integrated water management tool was conceived for use in research and policy making. De Lange *et al.* (2014) review the outcomes of this integrated water system analysis, as accounted for in the most recent updated of the Netherlands Hydrological Instrument (NHI). According to the outcomes of the research, surface water is managed based on surpluses or shortage, its salt content, and its temperature in an attempt to maximize efficiency for agriculture and consumer uses, preservation of natural resources, and other uses.

The NHI is made up of five models for different water domains and is a result of the work started in 2005, after pooling approximately one million euros a year to formulate a complete modeling system. Regional and national water management organizations across the country reached consensus for NHI in 2013, creating a modeling plan on a national, regional and local level. A main domain of the system is surface water management using surface water and sub-surface water modeling, aiming at optimizing distribution during drought. The individual models comprising the complete Instrument cover distribution of surface water, its flow and transport, water sub-catchments, saturated ground water, and soil and atmosphere water transfers. Together, they can produce plans for water allocation based on demand (from agriculture, drainage, seepage, etc.), models for water fluxes over time, and even the total salt load in kg/ha.

Apart from the advanced and comprehensive technical features of NHI, its success in establishing stakeholder participation is key to the success of water management in the Netherlands. Knowledge and data have been openly shared between local and national government, water companies, and water management organizations to create this cutting-edge tool. Its uses may span national policy development for management and distribution, to national and international research for sustainable water management and use. The lessons learned from such modeling constitute excellent examples for reform in areas of the US, such as Southern California. The NHI demonstrates the clear benefits of a participatory process based on information and expertise sharing, leading to a useful reset of resources and regional organization based on the water available. The results could then be implemented on a national level, ensuring access and sustainable use even in extended dry seasons, as the one California is currently experiencing.

Land Management During Conflict: A Case for Angola

Regions in the Angolan Highland plateau have been utilizing the Umbundu Land Use System (ULUS) throughout the country's conflict-ridden history. Addressing a gap in previous research, Delgado-Matas *et al.* (2015), examined the effect of conflicts in parallel with the dynamics of land use systems to understand how recovery develops and people adopt to the new conditions. Specifically, they looked at ULUS and how they can adapt to the social and economic changes in the communities that implement them. Their sustainability as a current system was also evaluated based on their effectiveness in supporting the economy in Angola.

Their experimental hypothesis was that conflict in the region drives the landscape's evolution via changes in land use and decision-making processes resulting in new governance structures. The study then, focused on a region of the Central Angolan Highlands that were considered essential to Angola's exports (beans, coffee, corn, etc.), even after 27 years of civil war in the region. The data analysis pulled from Portuguese colonial documents and officers' personal accounts of their time in Angola, national agricultural institutions, and NGO work. These were coupled with interviews of rural development officers and community leaders, and fieldwork with UN agencies and food NGOs.

Delgado-Matas and his colleagues revealed 9 consistent types of agricultural plots in ULUS. The analysis indicated that human-altered plots are the only nes with the potential to be both fertile and have good drainage, while each type of plot work was found to be gender-specific. Property and decision-making where shared among individuals or families and the land uses of said property were based on a catena structure as defined by the slope of the plots and human-nature interaction in them. Landscape and land use evaluation was performed under the multi-functionality lens (all demands are considered legitimate) and the civil war was found to have contributed to new functions of land use and changes in ownership from community to individual. During periods of peace, forest areas were found to decline, while fallows were good indicators of conflict periods. The comparison of different historical periods indicated that ULUS was generally more successful than colonial or other regimes, yet, interestingly, an attitude of dependence on humanitarian aid was observed post-independence.

These observations therefore indicate that the Umbundu system allows for development through conflict periods with partial pause of activities and continuation post-conflict. This shifts ownership to a more private setting (allows for better management and control in the long run), while fallows may increase land fertility.

Characterization of Cross-border Resource Management Organizations—the Case of the US, Canada and Mexico

International organizations and national or transboundary networks largely coordinate natural resource governance, scientific research management and policy. In the absence of coherent policy for cases like the U.S., Canada and Mexico, governments promote cross-border agreements with organizations for more successful collaboration of the actors involved. Stoett and Temby (2015) examine the role of intergovernmental institutions and transnational policy networks in the three states and propose a broad theoretical background and a functional description based on the nature of their activities and their internal governance.

Historically, it seems that a "fragmented" approach was taken to international resources issues in the area, with little trilateral agreement, even when faced with pressing issues of cross-boarder resources and important migratory species (like the gray whale). In the technocratic approach to managing resources, heavy industry appears to be challenged by environmental groups, while natural resource management is paralyzed in the absence of collective government action. With the realistic threat of climate change, Stoett and Temby urge for an extensive assessment of the organizations involved and for harvesting the value of scientific reports produced as authoritative work on cross border decisions.

From a Global Environment Politics perspective, this assessment might come in the form of effectiveness of the organization to solve the problem and affect others to contribute toward its management. A number of organizations possess the ability to produce the technocratic work to produce policies and establish networks of collaboration, with their authority being dependent on their autonomy from regional governments. These networks, in turn, inform policy on a governmental level, indicating that environmental management is a "multiagency and multistakeholder affair" and the organizations involved can be evaluated based on their "interactiveness" with other stakeholders in policy making.

In the 15 unique terrestrial ecoregions and the 19 marine ones shared between the US, Canada and Mexico, the authors find relatively well distributed typology for the organizations involved. A few like the Commission on Environmental Cooperation have higher influence and lead policy-related information sharing. Among others examined, the International Joint Commission, as a bilateral commission, exerts influence as a "participatory regulator" and "collaborative facilitator," when the Pacific Salmon Commission has regulatory authority.

The aim of this assessment consists of evaluating effectiveness of the management regime and answering the question: Is the urgency of action on climate change sufficient to alter policy networks and motivate policy-making for a more successful cross border collaboration?

Excessive Investments by Gold Mining Firms in Tanzania

Despite the history of gold extraction in Tanzania that started back in the late nineteenth century, and several protective policies early in the country's independence years, the current tax framework only yields about 2% of the total tax revenue. With 30% foreign direct investment, the losses for the government are extensive and unsustainable for the future of investments in the country. Shwilima and Konishi (2014) attempt to investigate the effect of governmental tax benefits on a firm's decisions and gold extraction patterns by designing an optimized model for all taxes affecting multinational firms and gold extraction. The foundations of their work lie in the sustainable development principle of intergenerational equity, the fact that a country ought to maintain stable per capita consumption rates over time for the benefit of future generations too. The principle is not achieved in Tanzania as research indicates that tax incentives minimize the income that could be invested in productive capital.

The current Tanzanian tax regime involves a profit and a royalty tax that remain constant over time, while benefits of delay of early tax payments favor the work of multinational firms. It allows for depreciation of assets and free profit repatriation among others. For the modeling they performed, the authors used a relatively outdated assumption of complete exhaustion of the resource after carefully considering the absence of significant changes in corporate policy according to recent studies.

During the time of the study, three international companies operated in mines in Tanzania: African Barrick Gold, AngloGold Ashanti, and Resolute Mining Limited. In the absence of corporate taxation in the first years of a company operating in Tanzania, the researchers found a declining rate of extraction for every company, following years of significantly higher than normally expected extraction. Part of that, they trace to mining development agreements set up with investors to ensure their consistent cooperation.

Some of the information they extract from their model equations include that the discounted marginal profit should be the same at all times and that unless the price of gold rises very fast, the extraction is pushed earlier in time, reducing the amount to be extracted in later years. The royalty tax, if set to remain stable throughout the extraction, was not found to affect the rate over time. The *ad volarem* nature of the Tanzanian tax could potentially dis-incentivize an earlier extraction. What seems to have an important effect however, is the profit tax that initially remains close to zero until the firms offset their losses, and then affects the gold extracted; extracting more minerals under the no tax regime seems, of course, very beneficial to the investors. The author's recommendations involve change of the full-depreciation allowances to a true-life regime and the signing of treaties that will allow for firm taxation at the profit transfers.

Deforestation Restrictions: Observations from Acre, Brazil

In light of climate change and species conservation efforts, "Reductions in Emissions from Deforestation and Degradation" (REDD) is becoming an increasingly important mechanism in conservation policy. Deforestation may affect water quality, forest services and local economies; its impact, as Pfaff *et al.* (2014) explore, depends on governance and location of protected forest areas. The researchers' work focuses on the forests of Acre, Brazil, which includes over 1 million hectares of protected areas. They evaluated deforestation in the periods of 2000–2004 and 2004–2008, and separated the protected forest areas in question in three categories (sustainable use, indigenous and integral). They used "remotely sensed pixel data" from the INPE (Instituto Nacional de Pesquisas Espaciais) to examine covered and cleared forestland and understand the potential effects of policy in a region.

Among other considerations, "protection of forests" involves the selection of an area, creation of a legal framework for conservation and complete management of the region, including monitoring technology implementation, creation of framework enforcement capacity and sustainable use of natural resources. Under law, a protected area is created following scientific studies and public consultations to taking into consideration and type of protection and its boundaries. Researchers focused on governance differences, since "sustainable use" usually targets populated areas, with higher threats to the forest and "integral protection" covers areas with few people. To avoid errors in the calculations, they only compared similar land using a "matching" approach instead of comparing total of forest and uncovered land.

Interestingly, between 2004 and 2008, the rates of deforestation fall for unprotected and for integral and indigenous protection, while they rise for locations of "sustainable use," making them face a higher threat. As a result, forest clearing prevented by such areas is also higher. Integral areas generally were found to have the lower rates. Areas declared protected before 2000 seem to be farther away from roads than unprotected but no differences in distances from cities or forest boundaries. The authors noted that restrictions on resource extraction are important, but enforcement efforts are still not successful; property-rights changes, on the other hand may affect monitoring and create negative results. They found that the approval of a local community for a protected area leads to higher monitor and accountability, thus making the process more successful and easy to implement. They concluded that less restrictive governance allows for deforestation, but produces a more successful outcome that classic strict protection. Some of these conclusions will hopefully be essential in the evaluation of resource allocation for REDD and determination of policy for deforestation control.

Conclusions

The decentralization of power and allocation of responsibility seem to become prominent elements in the evaluation of resources governance systems. In cases of land and water management, the communal or municipal level of administration, successfully maintain an efficient system and delivers results. In cases of deforestation and other land uses, great control might be needed to promote sustainability and accountability in a polycentric model of governance. As the international community redefines its need and potential for aid and establishes new tracks for development, scientifically-informed policy in the management of resources will increase the potential for success and better results catered to those in need.

References Cited

Bollig, M. & Schwieger, D., 2014. Fragmentation, Cooperation and Power: Institutional Dynamics in Natural Resource Governance in North-Western Namibia. Human Ecology 42, 167-181.

Delgado-Matas, C., Mola-Yudego, B., Gritten, D., Kiala-Kalusinga, D., Pukkala, T. 2015. Land use evolution and management under recurrent conflict conditions: Umbundu agroforestry system in the Angolan Highlands. Land Use Policy 42, 460-470.

Kininmonth, S., Bergsten, A., Bodin, Ö., 2015. Closing the collaborative gap: Aligning social and ecological connectivity for better management of interconnected wetlands. AMBIO 44, 138-148.

Kuzdas, C., Wiek, A., Warner, B., Vignola, R., Morataya, R., 2015. Integrated and participatory analysis of water governance regimes: The case of the Costa Rican dry tropics. World Development 66, 254-268.

Lange, W., Prinsen, G., Hoogewoud, J., Veldhuizen, A., Verkaik, J., Essink, G., Walsum, P., Delsman, J., Hunink, J., Massop, H., Kroon, T., 2014. An operational, multi-scale, multi-model system for consensus-based, integrated water management and policy analysis: The Netherlands Hydrological Instrument. Environmental Modelling & Software 59, 98-108.

Pfaff, A., Robalino, J., Lima, E., Sandoval, C., Herrera, L., 2014. Governance, Location and Avoided Deforestation from Protected Areas: Greater Restrictions Can Have Lower Impact, Due to Differences in Location. World Development 55, 7-20.

Schwilima, A. and Konishi, H. 2014. The impact of tax concessions on extraction of non-renewable resources: an application to gold mining in Tanzania. Journal of Natural Resources Policy Research 4, 221-232.

Stoett, P., Temby, O., 2015. Bilateral and Trilateral Natural Resource and Biodiversity Governance in North America: Organizations, Networks, and Inclusion. Review of Policy Research, 32: 1–18.

Economic Impacts and Future Expectations of Global Warming and Climate Change

Ali Siddiqui

Global Warming and climate change is a growing concern because of how it affects a nation's energy security, resource management, agricultural productivity, and even general food and safety. This chapter summarizes a variety of investigations led by different researchers about these very effects global warming and climate change have on particular nations as well as on a generic level. Each study's team of scientists not only investigates how a nation manages its energy and resources, but also advocates for different methods of resource management as a result of their investigation. Typically, the reasoning behind each team's advocacy of new styles of resource management stems largely from the economic consequences that will arise if the future expectations of global warming and climate change come to fruition. The first summary begins with a general investigation of China's general energy security, which has not improved despite 30 years of economic reform. This investigation provides a basic understanding of the large role economic policy has in shaping a nation's energy security and the need for adjustments by a nation in order to maintain energy security. Once this understanding is gained, the chapter then proceeds to look at a case study that investigates a specific example of when global warming caused large-scale economic consequences. This team of researchers uses data from the heat wave and drought combination that occurred in the US Corn Belt in 2012.

This particular study is, also, especially interesting because it then proceeds to use that same data and a modeling system to extend their case study to analyze the economic consequences if a similar weather situation were to arise in the future. The third summary is a case study, where the investigators attempt to analyze the agricultural setbacks that will arise in Brazil if global warming continues. One of the major results of this study is that global warming has negative impacts on land viability, which sets the stage for the next summary in this chapter. The fourth summary in this study investigates whether different agricultural techniques can help mitigate the negative impacts climate change has on crop production, but moves from Brazil back to the Midwest of the USA. The investigation led to some positive results, but the concern about global warming and climate change is still present. The fifth study demonstrates why the concern is still present because it investigates how if future expectations of global warming are true, then certain species of arable-weed species may become endangered. This study relates to the previous two because of the large role arable-weed species play in increasing the biodiversity of the land, which is beneficial for agricultural purposes. The sixth summary is of a study that attempts to assess the impacts climate change will have on food and water safety across North America. The interesting aspect of this investigation was the researcher's modelling program that allowed them to assess impacts of different degrees of climate change based on different expectations of the rate of global warming and climate change. The seventh summary returns back to the theme of economic policies and energy security. The summary is of a study that looks at the *Jatropha curcas* as a potential biofuel, which Mexico has actually been subsidizing and the economic consequences of that decision. The final eighth summary is of a study that investigates how European expectations of wind energy as an alternative energy source and climate change will affect European energy policy. This summary was chosen last because not only does it represent how expectations of climate change can adjust an entire continent's energy policy, but also because it shows how accounting for that expectation can help maximize an energy policy.

Chinese Energy Security

According to Yao and Chang (2015), China's energy security has not improved over 30 years of reform. The authors aimed to understand why in a qualitative fashion, finding a critical component is China's macroeconomic reform. By analyzing China's energy security in this fashion, these researchers hope to broaden the perspective on the way developing economies in transition conduct research on their energy security.

The researchers have defined energy security as availability of energy resources, applicability of technology, acceptability by society, and affordability of energy prices. They demonstrate how, after the Third Plenary Session of the 11th Chinese Communist Party Central Committee session, the focus shifted to decentralizing and distributing economic power to the lower levels of the economy, which allowed for reform in the microeconomic level of the economy. Incentives for farmers to produce more led to better efficiency in the agricultural sector of China's economy, which led to a surplus of labor from that sector that shifted into the coal mining sector that was also then beginning to be supplemented by the state. When these same incentive policies were put to the urban area, China began to move in a direction that reflected a market economy. This continued pro-local autonomy eventually led to the financial constraint in China's power industry to be removed and the electric supply to increase. This relaxed economic environment of the 1990s led China to speed up reform in order to legitimize the "socialist market economy" including government structure, national fiscal and financial systems, and state-owned enterprise reforms.

These reforms also led to the decentralization of leadership and thus ability to regulate in any administrative setting the energy sector of the economy. Therefore, macroeconomic reforms that resulted as a reaction to the increased growth and production that in turn resulted from microeconomic reform led to a mismanagement of energy policy that hindered China's ability to manage its energy security. The view of these researchers is that new investment into the sector by private firms, favorable tax policies for saving energy, and introducing foreign competition could alleviate China's issues.

Weather Extreme Impacts Maize Yield and Food Security

The US is the leading maize exporter of the world, but in 2012 The United States Department of Agriculture (USDA) announced that maize production had been reduced by 13%. The reduction in maize production occurred due to the combination of a heat wave and severe drought that occurred from May till August 2012 in the US Corn Belt. While various definitions of a heat wave is not consistent with respect to temperature threshold, metric, and number of days, heat waves are typically understood as prolonged periods of extreme heat. The heat wave and drought combination, however, did not affect the quality of the production according to the USDA.

Chung *et al.* (2014) are interested in the interplay between heat waves and global crop production and use the heat wave of 2012 in the US as a case study. Their ultimate goal is to "determine the biophysical impact of weather extremes on maize production in the USA] and assess the secondary effects of such weather extremes on world maize prices, production, consumption, and trade."

The investigators used a process-based crop model known as CERES-Maize to examine the implications of a similar extreme weather event situation in the US in 2050 and delineate its results as well. The results of their study showed that global prices remained largely unaffected in 2012 by the US's decreased maize production because Brazil had record maize harvests, China had stable maize production, and the EU had increased production. However, as the heat wave and drought continued and the EU suffered a summer drought, global prices rose. In August 2012 yellow maize reached a record high of $331 per ton. The results of their extension of the case study using the modelling program suggested that maize yields would be reduced by 29%. Overall, the investigators concluded that in both the case of 2012 and 2050, the countries that are impacted the most, however, are those that have a high import ratio or countries too poor to purchase the imports at the increased global price like those in Sub-Saharan Africa.

Climate Change Effects on Different Brazilian Farming Scenarios

The International Panel on Climate Change has indicated that rising temperatures in Brazil, due to its location in tropical and subtropical areas, will cause setbacks in its agricultural industry. Filho and Moraes (2015) wanted to understand not only the relationship between climate change and agriculture production, but also the greater impact of climate change on the Brazilian economy.

They take into account the benefits that would come to sugarcane and cassava, as well as look at the effects on the Brazilian labor market, by attempting to examine how climate change affects how income is distributed. Their model is based on three distinct databases that are the Brazilian-output and input table, Brazilian National House survey, and the Brazilian expenditure survey, and incorporate previous models. One noted flaw in their design is that the base framework for their model is data collected in 2005. Therefore, adaptations to climate change that may have taken place after the projections had been forecasted-for example; effects on crop viability-may not have been included leaving room for improvement in the future.

One of their conclusions from taking samples across the Brazilian region is that despite the increasing temperatures, the southeastern region is affected the least. The reasoning is due to the high volume production of sugar cane, a crop with high resistivity typically unaffected by a loss of land viability. Overall, the authors found that long-term effects on the economy are not large, however, on the regional level; the poorer regions especially in the northeast, where agriculture is most pertinent to their economy suffer large consequences due to the climate change's effect on land viability etc. The authors suggest to truly understand climate change and its far-reaching implications into the economy more research is needed.

Climate-Smart Agriculture and Biophysical Consequences in the Midwest

Climate-smart agricultural techniques are agronomical practices that help alleviate the consequences that climate change has on agriculture. Agronomical practices are related to soil management and production of field crops. Currently, in the Midwestern US different climate-smart techniques have been advocated in order to increase crop production. These include utilizing different crop cultivars in order to reap the benefits of earlier planting dates and a longer growing season and no-till agriculture in order to reduce soil emissions and maintain soil moisture. Bagley, Miller, and Bernacchi (2015) using observational data and an agroecosystem model that uses future temperature and CO_2 concentrations determine the effectiveness of climate-smart techniques and their biophysical impacts.

The first scenario of interest uses predicted future temperatures and CO_2 emissions from their model and results in spring snowfall lessening and a warmer autumn, which led to crops such as maize emerging 6–11 days and finishing maturation 15–30 days earlier than the current season. The potential biophysical impact of this result was a decrease in water availability as crops mature due to the increased time that the crops are actively transpiring. The cultivars–plants produced by selective breeding–used in order to take advantage of this extended growing season had short growing seasons and were compared to cultivars with long growing seasons. The two maize cultivars had almost identical simulations until the cultivar with the shorter growing season reached day 200 after planting and was ready to be harvested, unlike the longer growing season maize cultivar.

The second climate-smart agricultural technique was the use of no-till agriculture. No-till agriculture has been advocated as a climate-smart agricultural technique because it reduces soil temperatures and retains soil moisture, as well as because the debris left over is beneficial to the soil. The biophysical impacts of this technique were that in the spring and autumn month's soil moisture was at its highest saturation leading to overall better production in yield. Overall, the conclusions drawn by the results of both scenarios were that the no-till agriculture led to better overall biophysical impacts than using cultivars to take advantage of the extended growing season.

The Future of Arable Weed Species with Climate Change

Rühl et al. (2015) writing in *Biological Conservation* attempted to determine whether predicted changes in temperature as caused by global warming would be detrimental to the survival of endangered arable weed species, which play an important role in increasing biodiversity. The results of the study were multiple. The authors concluded that endangered arable weed species germinated significantly less than the common arable weed species under increased temperature conditions and preferred lower optimal germination temperatures ($24\pm3.5°C$) than common arable weed species ($31\pm0.5°C$).

The researchers also found that common arable weed species tended to have greater flexibility when it came to changes in water potential for the seeds. The conditions for endangered arable weed species to germinate successfully were therefore much narrower than the common arable weed species. In fact, the scientists even discussed different germination strategies employed by common arable weed species to avoid complete failure of all seedlings' germination.

The importance of temperature change on species' ability to germinate was extended to incorporate long-term consequences. The authors showed how flexibility to temperature change for germination by some species has allowed those species to maintain population size, whereas other species have undergone a reduction.

The arable weed species have an unfavorable genetic structure comprising of low genetic diversity and high differentiation between populations, which leaves them to an unpromising future with predicted climate change. The evolutionary response of the common arable weed species to become more flexible to rapidly changing environments as a consequence of coexisting with crops was suggested a possible explanation for the difference between the germination success of the endangered arable weed species that does not coexist with crops, and the common weed arable species.

The paper ended by drawing on the larger question as to how

this complex interplay between climate change and the survival of endangered species with unfavorable genetic structures will play out.

Risk Assessment of Climate Change Impacts on Food and Water Safety

According to the background research done by Ben A. Smith *et al.* temperatures and total annual precipitation across most of North America are expected to rise. They suggest that these expected changes would impact areas negatively especially if those areas are developing and population is increasing. The negative impacts arise from how food and water contamination can be increased by the longer survival of both old and new pathogens as well as the extended peak season for many microbial diseases. An example they provided was of *Salmonella* infections, which correlated with rising global air temperatures in most continents besides Europe, negated the rise in infections by using human intervention through public health. This intervention was implemented once a quantities microbial risk assessment (QMRA) model was utilized. QMRA models are typically developed assuming historical/static climate conditions. These researchers suggest, however, that adding climate change factors in food/water safety QRMA models make them increasingly complex due to the varying range of relevancy of variables. According to the authors, a framework is necessary in order to better understand the large data alongside the QRMA model elements so as to assess the potential impacts of changing climate variables on public health.

The framework proposed is subdivided into three categories: knowledge synthesis, data storage and access, and stochastic QMRA modeling. These subdivisions allow for identifying and establishing relationships between climate variables and safety elements, applying data sets across risk models, and integrating data in an effective mathematical fashion. The researchers have developed three case studies in different regions of Canada and analyzed climate impacts (current and projected), hazards, commodity production and processing, and population demographics and behaviors. They analyzed ochratoxin A (in wheat grown in Saskatchewan), *Vibrio parahaemolyticus* (oysters in coastal British Columbia), and *Giardia* in drinking water of a generic northern community. The researchers then subjected each of these case studies to a food/water safety model, where they attempted to understand the projected change in concentrations of the ochratoxin A, *Vibrio parahaemolyticus*, and *Giardia*.

To ensure the effectiveness of their models, data extracted from primary literature, government reports, and existing databases were categorized in case-specific and crosscutting databases. Case specific databases were analyzed as point estimates directly put into their Analytica software-modeling program. Cross cutting data like regional air and water temperatures were used to inform their QMRA models. The Analytica studies were subjected to multiple simulations sometimes over 1000 that changed slightly each time to represent different situations. All these different representations therefore served as potential data points for the overall framework, which could then be applicable in many different situations.

Ultimately, the researchers created a framework which provides a platform to asses varying climate factors, their risk, and provide mitigation options to reduce food and waterborne diseases using risk models.

Jatropha curcas the Sustainable Biofuel?

Biofuels are a controversial energy source; however, there has been strong interest in these biofuels by both developing and industrialized countries. Some countries have even financially incentivized the production of these crops. A promising crop called *Jatropha curcas* has recently been viewed as a sustainable option for biodiesel production. This inedible plant has typically been used for soap production, medicinal purposes, and even as way to demarcate property boundaries as a living fence. Its high resistance to droughts and pests, inability to be eaten by cattle, short gestation period, and versatility of products are qualities that have been cited by many researchers as advantages for small farming communities, who might be able to use *Jatropha* to increase their employment, increase revenue, and increase energy self-reliance.

Although researchers hypothesize advantages for communities adopting this crop, some communities that have have suffered real negative consequences. The study by Iria Soto *et al.* (2015) tries to understand whether the state of Chiapas, Mexico, subsidies to farmers growing *Jatropha would* result in an overall better economic climate for the entire community.

The results revealed many economic effects caused by Chiapas's subsidy program. One trend that arose was that the communities that most often received a greater portion of subsidies were those that were generally larger and more inter-connected with other communities. Another trend noticed was that within the communities that adopted the new crop, the subsidies tended to be allocated to households that were more advanced either through better technology, access to information, and/or more resources. An implication of these trends was that the way the Chiapas allocated subsidies to communities led to the lesser distribution of government resources to poorer communities as well as less resourceful and smaller households.

Chiapas used an extensive network of agents to reach out to rural communities in order to inform them of this government program and to persuade farmers to adopt *Jatropha curcas*. However, the state also incentivized these agents to get most communities to adopt these new practices through financial means. This incentive program may have led agents to reach out to communities they believed could sustain the risks of these new practices, which may have led to the trends outlined earlier. Overall, a different strategy needs to be implemented by the state in order to target rural farmers, who may not be as relatively wealthy or advanced as farmers in larger communities.

Assessing Future Changes in Potential Generation of Wind-Power over Europe

Although wind power currently makes up about 3% of the world's generated electricity, wind power provides in Ireland, Spain, Portugal, and Denmark 15-30% of national electricity. Near-surface winds are typically those that power wind turbines and can be affected by climate change. Tobin *et al.* (2014) are interested in assessing how climate change can affect the potential wind power generation over all of Europe and effective power production from operational wind farms at the end of 2012 and 2050. To reduce uncertainty stemming from model formulations, these investigators used multi-model investigations. For example, they used a multi-model ensemble of 15 regional climate projections from the ENSEMBLES project, downscaling 6 Global Climate Models with 10 Regional Climate Models to gauge effects on potential wind power generation.

Overall their results concluded that future changes in wind power potential are weak or non-significant over a large part of Europe. Future changes in wind power production are computed as the difference between the mean values 1971–2000 present period and the 2031–2060 and 2071–2100 future periods. They are considered robust when the mean difference between the values is at a 95% confidence interval for statistical significance and that 12 out of their 15 models agree with them.

The model showed that wind power potential will stay within ±15 to ±20% by mid and late century, and determined that Northern Europe would see a robust increase in wind power potential, while the Mediterranean will see a decrease. Their main conclusion was that climate change would not severely affect wind power potential either adversely or beneficially, however, planning wind farms by accounting for climate change effects in particular regions may help to maximize wind power development in Europe. These conclusions do, however, have to be understood with the knowledge that these models do not take into account the full range of uncertainty that stems from model formulations.

Conclusions

These summaries have all described how future expectations of climate change are that global warming will continue to increase and that these will cause negative consequences for many nation's economies unless these nations attempt to mitigate the problem through policy that accounts for that temperature increase.

References Cited

Bagley, J. E., Miller, J., Bernacchi, C. J. 2015. Biophysical impacts of climate-smart agriculture in the Midwest United States. Plant, Cell and Environment, 1-18. DOI: 10.1111/pce.12485.

Chung, U., Gbegbelegbe, S., Shiferaw, B., Roberston, R., Yun, J., Tesfaye, K., Hoogenboom, G., Sonder, K. 2015.Modeling the effect of a heat wave on maize production in the USA and its implications on food security in the developing world. Weather and Climate Extremes 5, 67-77.

Filho, J., Moraes, G. 2015. Climate change, agriculture, and economic effects on different regions of Brazil. Environment and Developmental Economics 20, 37-56.

Rühl, Theresa A., Eckstein, Lutz R., Otte, Annette, Donath, Tobias W. 2015., "Future challenge for endangered arable weed species facing global warming: Low temperature optima and narrow moisture requirements." Biological Conservation 182, 262-269.

Smith, Ben. Ruthman, Todd., Sparling, Erik. Auld, Heather, Corner, Neil, Young, Ian, Lammerding, Anna M., Fazil, Aamir. 2014. "A risk modeling framework to evaluate the impacts of climate change and adaption on food and water safety." Food Research International 61, 1-8. Soto, I., Achten, W., Muys, B., Mathjis, E. 2015. Who benefits from energy policy incentives? The case of Jatropha adoption by smallholders in Mexico. Energy Policy 79, 37-47.

Tobin, I., Vautard, R., Balog, I., Bréon, F., Jerez, S., Ruti P., Thais, F., Vrac M., Yiou, P. 2014. Assessing climate change impacts on European wind energy from ENSEMBLES high-resolution climate projections. http://link.springer.com/article/10.1007/s10584-014-1291-0/fulltext.html

Yao, L., Chang, Y. 2015. "Shaping China's energy security: The impact of domestic reforms." Energy Policy 77. 131-139.

About the Authors

The authors of this book are students at the Claremont Colleges. The book is a work product of Biology 159: Natural Resources Management taught by Emil Morhardt in the W.M. Keck Science Department of Claremont McKenna, Pitzer, and Scripps Colleges. Each student picked a topic, did a literature search, and selected eight papers written within the past year that exemplified the state of the science.

Their task was to write journalistic summaries capturing the essence of the papers but eschewing technical terms to the extent possible—to become, in effect, science writers. The summaries were due weekly and were returned with editorial comments shortly thereafter. The chapters are compilations of the individual summaries with additional introductory material and a brief conclusion.

The editor is Roberts Professor of Environmental Biology at Claremont McKenna, Pitzer, and Scripps colleges. He remembers how difficult it is to learn to write and appreciates the professionalism shown by these students.

Index

abdominal profile index (API), 121
acidification, 172, 183, 184, 197
Acropora millepora, 166, 175
Aedes japonicus japonicus, 281
aerosols, 49
Africa, 269, 277, 331, 349
agricultural intensification, 313
agriculture, 54, 146, 189, 190, 192, 193, 194, 195, 196, 198, 199, 200, 201, 203, 207, 212, 217, 219, 220, 286, 311, 313, 317, 323, 337, 349, 350, 356
albedo, 52, 319
Anser albifrons flavirostris, 120
Antarctica, 313
Antelope Valley, 45
Antelope Valley Solar Ranch, 45
aquaculture, 185, 309
Archontophoenix cunninghamiana, 144
Arctic, 146, 156, 321, 322
Ardabil, 37
Argentina, 19, 279, 280
Australia, 15, 104, 107, 114, 158, 173, 174, 179, 187, 228, 237, 296
Barcelona, 263, 264, 273
BaseTrace, 74
battery, 13, 14, 15, 16, 17, 18, 19, 20, 21, 22, 23, 24, 25, 26, 31, 41
biodiesel, 353
biodiversity, 38, 46, 96, 131, 132, 134, 135, 137, 139, 140, 143, 146, 148, 150, 155, 169, 173, 177, 178, 181, 182, 183, 184, 186, 189, 207, 209, 210, 215, 216, 218, 220, 222, 223,

225, 230, 231, 234, 235, 236, 237, 317, 329, 346, 351
bioerosion, 163
bird hazard index (BHI), 48
bleaching, 159, 161, 162, 163, 164, 165, 168, 171, 174
Brazil, 227, 342, 346, 348, 349, 356
brown, 122, 265
brownfield, 264, 265, 266, 273
bumphead parrotfish, 182
Bureau of Economic Geology (BEG), 84
butterfly, 122, 123
C_3 plants, 166, 167
calcium carbonate ($CaCO_3$), 158
California Current Large Marine Ecosystem (CCLME), 183
CalTech Jet Propulsion Laboratory, 22
Canada, 221, 339, 340, 352
cap and trade, 248, 256
Carbon Capture and Storage (CCS), 29, 30, 42, 66, 70, 108, 114
Carbon Capture Storage (CCS), 28
carbon intensity (CI), 270
carbon sequestration, 209, 214, 263, 264
carbon sink, 211, 221, 321
Caribbean, 160, 163, 164
Changhua, 234, 237
China, 21, 31, 51, 206, 212, 213, 224, 234, 237, 302, 303, 315, 318, 330, 345, 347, 348, 356
Chironommus riparius, 320
chronic kidney disease (CKD), 285, 286
climate forcing, 68
climate model, 142, 169, 191, 246, 265, 278, 303

CO, 29, 140

coal, 27, 28, 29, 30, 35, 42, 46, 51, 58, 68, 78, 79, 83, 85, 106, 112, 252, 347

Colias meadii, 123

Community-Based Natural Resource Management (CBNRM), 332, 333

computational fluid dynamic (CFD) model, 91

Concentrated Solar Power (CSP), 49, 50

conventional tillage (CT), 195

coral, 158, 159, 160, 161, 162, 163, 164, 165, 166, 167, 168, 169, 170, 171, 172, 173, 175, 179, 180, 182, 187

coral bleaching, 158, 159, 161, 162, 167, 171, 172

corridor, 40

Costa Rica, 124, 331, 335, 344

cryoprotectant, 120

CubeSat, 21

current energy converter, 89

cyclone, 179

cyclones, 179

Daphnia magna, 320

decomposition, 235, 247, 248, 258, 321

deforestation, 117, 203, 206, 207, 216, 226, 227, 228, 229, 230, 231, 233, 234, 236, 237, 305, 342, 343

Delft3D-FLOW model, 90

Dermochelys coriacea, 124, 127

dinoflagellate, 145

dissolved oxygen (DO), 322

Driver-Pressure-State-Impact-Response (DPSIR) model, 323

Drospohila, 148

Ebro, 317, 319, 320, 329

ectotherm, 122

El Niño, 296, 297, 300, 301

El Niño Southern Oscillation (ENSO), 296, 297, 300, 301

El Niño Southern Oscillation, ENSO, 296, 297, 300, 301

electrolyzer, 19, 31

Electrophorus electricus., 24

Endangered Species Act (ESA)., 130, 131, 135

endemic, 131, 144, 234, 295, 301

endemism, 234

energy efficiency, 16, 101, 249, 254

engine, 17, 18

ENVLinkages, 112

erosion, 45, 46, 50, 54, 79, 89, 94, 95, 162, 163, 195, 200, 210, 214, 304

ethane, 59

ethanol, 54, 55

Europe, 53, 54, 90, 118, 145, 281, 290, 291, 324, 352, 354, 355

European Observation Network, Territorial Development and Cohesion (EPSON), 290, 291

European Union Emissions Trading Scheme (EU ETS), 247

European Water Framework Directive (WFD), 320

eutrophication, 145, 197

evapotranspiration, 189, 191, 192, 226

excess winter deaths (EWD), 307

extinction, 119, 129, 131, 132, 133, 137, 138, 153, 154, 156, 170, 173, 174, 229, 235, 309, 311

fertilization, 190

fertilizer, 145, 195

fire, 208, 225, 226, 295, 304, 305, 306, 307

FLOWBEC Seabed platform, 98

flux, 61, 317, 321, 322, 330, 337

flywheel, 19, 20

folpet, 198

forest, 79, 203, 204, 205, 206, 207, 208, 209, 211, 212, 213, 214, 215, 216, 217, 218, 220, 221, 222, 223, 224, 225, 226, 227, 228, 229, 230, 231, 232, 233, 234, 236, 237, 338, 342

FracEnsure, 74

fracturing fluid, 58, 60, 61, 86

Fukushima Daiichi, 103, 104, 107, 108, 110, 112, 113, 114

gasoline, 268, 269

Global Energy Assessment (GEA), 268

global gridded crop models (GGCM), 192

global warming potential (GWP), 197

grazing, 163, 164, 166, 190

Great Barrier Reef, GBR, 157, 158, 166, 172, 173, 174
Guadalete River, 322, 329
Heliotropium L. sect. *Cochranea*, 154
Hothaps program, 283
human modified tropical landscape (HMTL), 230
hurricane, 163
hybrid, 17, 18, 22, 23, 25, 26, 30, 106, 327
hybrid vehicle, 17
hydraulic fracturing, 57, 58, 59, 60, 61, 62, 64, 65, 66, 67, 68, 69, 70, 71, 72, 73, 74, 75, 76, 77, 78, 79, 80, 81, 83, 84, 85, 86
hydroclimate, 262
hydropower, 106, 135
ice sheet, 245, 246
Ice sheets, 258
Iceland, 120, 121
India, 207, 223, 284, 285, 286, 287
Infectious gastroenteritis, 296
Intergovernmental Panel on Climate Change (IPCC), 118, 142, 150, 203, 223, 278
International Association of Public Transport (UITP), 268
invasive species, 130, 131, 142, 143, 144, 235, 282
Iran, 28, 36, 38
isotope, 59
i-Tree Eco model, 264
Ivanpah, 45
Ivanpah Solar Power Facility, 45
Ixodes ricinus, 117, 118, 126
Kano, Nigeria, 266, 267, 273
Karenia, 145, 146
kerogen, 71
keystone species, 179
La Salle County, Texas, 79, 86
Lacandona, 230, 236
landfills, 82
Lasiommata megera, 122
leaf area index (LAI), 226
Levelized Cost of Energy (LCOE), 29
life cycle assessment (LCA), 201, 311
life cycle impact assessment (LCIA), 197

Lyme borreliosis, 118
macroalgae, 159, 162
maize, 192, 348, 350, 356
malaria, 275, 277, 278, 292, 296, 302
manure, 82, 195
Marcellus shale, 58, 62, 64, 67, 80, 83
Marcellus shale formation, 83
marine protected area (MPA), 177, 180, 184, 185, 187
marine renewable energy (MRE), 92, 93, 102
marine reserves, 179, 184, 185, 186, 187
Mekong Delta Basin, 277, 288, 289, 293
methane (CH_4), 58, 59, 69, 72, 76, 81, 82, 83, 84, 85, 86, 321, 322, 323, 329, 330
Michigan, 120
Mississippi River delta, 317, 326, 329
Mojave, 45
Mojave Solar Project, 45
Montastaea cavernosa, 164
mountain hemlock (*Tsuga mertensiana*), 152
mussel, 125
Mycoplasma pneumonia, 295, 300, 315
Mytilus edulis, 125
Namibia, 331, 332, 333, 343
net ecosystem carbon balance (NECB), 321, 322
Netherlands, 331, 336, 337, 344
New York, 9, 58, 67, 127
New Zealand, 143
Nile, 324, 325, 329
nitrogen, 25, 26, 39, 40, 43, 55, 69, 70, 86, 114, 125, 126, 127, 139, 145, 162, 173, 174, 187, 192, 195, 201, 223, 224, 236, 256, 258, 292, 322, 329, 350
nitrogen, N, 25, 26, 39, 40, 43, 55, 69, 70, 86, 114, 125, 126, 127, 139, 145, 173, 174, 187, 201, 223, 224, 236, 258, 292, 329, 350
nitrous oxide (N_2O), 322, 323, 329
Nitzschia paleax, 320
No-take marine reserve (NTMR), 178, 179
no-till (NT) agriculture, 350

Index

Nucella lapillus, 125
Nyctalus leisleri, 38
ocean acidification, 158, 170, 172, 179, 184, 186
Ocean Margin Ecosystems Group for Acidification Studies (OMEGAS), 183, 184
Odontomachus chelifer, 227
oil-shale, 77
once-through fuel cycle (OTC), 107
Oryza sativa (rice), 192, 195, 325
ozone, O_3, 81, 197, 312, 313, 314
Pachycondyla striata, 227
parasite, 276, 278, 279
parasites, 282
particulate organic carbon, POC, 322
peatlands, 321, 322, 330
Pennsylvania, 58, 62, 64, 67, 70, 80, 83, 86
permafrost, 321, 322
Peru, 231, 236
pesticide, 144, 196, 198
Pheidole, 227
phenology, 126
photosynthesis, 168, 191, 314
phytoplankton, 314
pika, 143, 149, 150, 156
Pinus sylvestris (Scots pine), 151
plug-in, 20, 21
Portugal, 354
primary productivity, 219, 321
proppant, 76
Prorocentrum, 145, 146
proton exchange membrane electrolysis cell (PEMEC), 34
Pseudokirschneriella subcapitata, 320
Psidium guajava, 144
rainforest, 228, 229, 230, 236, 237
reflectometer, 50
respiration, 321
Rhine, 336
Rhodnius prolixus, 279
Rhopalostylis sapid, 144
Río Potrero, 335
risk assessment, 352
Rocky Mountains, 123, 152, 155
runoff, 54, 145, 195, 268, 323, 327, 328

Russia, 276, 321
Salamander woodland, 119
salinity, 141, 145, 147, 170, 183, 194, 322
salinization, 325
salmon, 81
San Onofre Nuclear Generating Station, 104, 110, 111
Schefflera actinophylla, 144
schistosomiasis, 302
sea surface salinity, SSS, 145
sea surface temperature (SST), 180, 297
sea-level rise (SLR), 139, 170, 246, 258, 324, 325, 329
sequestration, 86, 199, 209, 252, 264
shale oil, 78
sheep, 82
Solenopsis, 227
solid oxide electrolysis cell (SOEC), 34
Spatial distribution models (SDMs), 39, 142, 144, 153, 154, 181, 182
species distribution model (SDM)., 39, 141, 142, 144, 153, 154, 181, 182
stomate, 191
stream flow, 81, 194
sugar cane, 349
supercapacitor, 20, 22
Sweden, 321, 331, 334
Symbiodinium, 164, 165, 166, 174
Tabriz, 36, 37, 38, 43
tar sands, 77
terbuthylazine, 198
terrestrial carbon flux (TCF), 321
Thailand, 28, 35, 36, 42
ticks, 118
tidal energy, 87, 88, 89, 90, 91, 94, 95, 96, 99, 101
tillage, 195, 200
Triatoma infestans, 279
trophic, 138, 147
tropical rain, 138, 209, 234, 236
Trypanosoma cruzi, 279
turbine, 13, 16, 17, 26, 28, 37, 43, 88, 89, 91, 92, 95, 96, 97, 98, 99, 100
twice-through nuclear fuel cycle (TTC), 107

upwelling, 183
urban, 45, 52, 56, 135, 136, 213, 259, 260, 261, 262, 263, 264, 265, 266, 267, 268, 269, 270, 272, 273, 274, 276, 284, 288, 318, 319, 320, 347
Urban Heat Island (UHI)., 52, 260, 261, 284
US Department of Agriculture (USDA), 47, 348
utility-scale solar energy (USSE), 46, 47
UV radiation, 295, 313, 314

vegetation, 46, 47, 54, 55, 199, 204, 206, 209, 211, 213, 220, 221, 225, 229, 235, 260, 261, 263, 264, 265, 266, 313, 321, 326
very large marine protected areas (VLMPA), 185
vulture, 133, 134
white bark pine (*Pinus albicaulis*), 152
windpower, 22
Wisconsin, 81
wood frog (*Rana sylvatica*), 120
World Bank (WB), 268